Uncovering the
Secrets *of* Time
and Number

Uncovering the Secrets *of* Time and Number

Finding Patterns and Rhythms in Everyday Life

WOLFGANG HELD

Floris
Books

Translated by Matthew Barton

Part 1 first published in German as *Der siebenfache
Flügelschag der Seele* in 2004 by Verlag Freies Geistesleben
Part 2 taken from *Vier Minuten Sternenzeit*, first published
in German in 2006 by Verlag Freies Geistesleben
Part 3 first published in German as *Alles ist Zahl: Was uns die Zahlen 1 bis 31
erzählen* in 2011 by Verlag Freies Geistesleben
© 2004, 2006, 2011 Verlag Freies Geistesleben & Urachhaus GmbH, Stuttgart

Previously published separately as *Rhythms of the Week* in 2011,
and *The Quality of Numbers 1 to 31* in 2012 by Floris Books
This combined English edition © 2020 Floris Books

British CIP Data available
ISBN 978-178250-664-5
Printed by Lightning Source

Contents

Foreword

We have conquered space. Every last bit of land has been measured and even the furthest corners of the globe have been mapped and can be reached by some means of transport. Today's voyages of discovery no longer take place in physical space but in the dimension that still often confounds us: time. In the twenty-first century this is the terra incognita, the unknown realm. Whether in the temporal configurations of our biography or the rhythmic effect of times of day and seasons of the year on our psyche and physical organism, we are finding increasingly that time does not flow in a uniform way but is, rather, a fabric woven of the most varying, changing qualities. It is high time, perhaps, that we tried to notice it more, to understand and incorporate it into our lives with greater awareness.

Few things have such a decisive affect on our positive personal development as our individual relationship with time. A hundred years ago, the Austrian poet Ödön von Horváth lamented that he rarely found time for anything. If this makes us smile today it is because we still relate to it, in fact more so than ever.

In the twentieth century, time-management advisers were largely the preserve of commerce and business, but today, increasingly, we are all faced by the vital question of our relationship to time. In the eighties, Gunther Hildebrandt, a pioneer in the research of biological rhythms, stated that 'Every person today is caught in a chronobiological time conflict.' Our modern lives are such that we are frequently at odds with our organic rhythms and therefore continually weaken our own vitality. Of course none of us can avoid the hectic pace of things today. The question is therefore not how to avoid this life-sapping way of living but rather how to cope better with the arrhythmia of modern life and ultimately organize it in a masterful and musical way.

Rhythm is always the best way of structuring our time. To develop new rhythms in our own lives, it is worth exploring and learning about the many existing rhythms and inner qualities at work in temporal processes. By this means we can gain a relationship to time and, instead of battling against it, make it into our tool and ally.

The voyage of exploration into the structure of the seven-day week described in Part One of this book aims, like any good guidebook, to facilitate your own discoveries, helping to get a purchase on this fleeting element so that it can serve you better in daily life.

Part Two offers some insights into other rhythms of life, to help acquaint you better with time and its possibilities. It encourages you to explore some previously unknown rhythms and use them creatively in your life.

Wolfgang Held

1 | Living with the Rhythms of the Week

The Soul's Sevenfold Wingbeat

What underlies the
seven-day week?

Water and soul are related

Comparisons are often made between the human psyche and
the sea. The ocean is as broad and unfathomable as the human
soul, but they have something else in common too – movement.
While it seems as if water is always set in motion by an external
cause such as the wind, a closer look can tell us that movement is
intrinsic to the nature of water itself. A simple test can illustrate
this: if you let a trickle of water run down a sloping sheet of glass,
the water will of course follow the slope, heeding gravity, but at
the same time will also seek its own, winding and continually
changing path. Water has a tendency to pulsate and move:
an inner vivacity which, over long periods, gives rise to those
wonderfully curving river meanders that engrave themselves
into the landscape – as long as water is not compelled to flow in
straight, man-made channels. Even when still, water is usually
in movement or expecting movement: a gentle breeze is enough,
the flap of a fin, and immediately water responds with rhythmic

waves. It does, though, need some kind of external stimulus to begin to stir.

The sovereign soul

The human soul is related to these qualities. Every sensory impression or external stimulus can set our soul in motion, evoking inner response in the form of feelings, thoughts or even actions. It is not for nothing that we say someone is 'cold' if they do not respond to outward occurrences with interest and involvement. They appear, then, to be like frozen water that likewise loses its 'love of movement', its receptive sensitivity.

But whereas water is reliant on outer stimulus to begin pulsing and swirling, we are capable of initiating movement without external cause, entirely out of ourselves. At such times we can experience everything external as distraction, and become aware of hidden inner movement within us. This capacity is wonderfully depicted in the figure of Baron Munchhausen, who pulls himself up by his own tuft of hair to free himself from a swamp. The soul can find its own impetus for movement and development, and there is probably no religion in the world that does not acknowledge this capacity as the soul's sovereign or regal nature. Language itself contains much wisdom: we speak both of raising and deepening in relation to the ego's self-initiating power of impetus, whether through reflection, meditation or prayer. At such times the soul resembles a calm mountain lake, reflecting the heavens yet also allowing us to see into its depths.

The soul swings back and forth between two extremes – the stronger it does so the better – finding stimulus and renewal in the outer world then returning home to be entirely centred in itself. The more we are able to withdraw into our own thoughts with a good book, a thought-provoking saying or any other content of reflection – even for just a few minutes each day – the more interest and participation we can call upon when meeting others, engaging with new ideas or going out into nature.

Those who practise such things will notice that on some days of the week they succeed less well than on others. For instance, it seems harder to focus on oneself on Wednesdays, while on Thursdays it is easier to make decisions and meet challenges. Why is this?

The weekly rhythm

The week has a remarkable rhythm that does not precisely accord either with the month or the year: a particular date will always fall on different days. In purely commercial terms, the week is the least practical aspect of our temporal divisions into second, minute, hour, day, week, month and year. Yet most of humanity keeps faith with this sevenfold rhythm. Seen from the point of view of astronomy, the week is also an exception. Whereas the month is derived from the moon, and the day and the year from the sun's circuit, the week does not accord with any temporal division exemplified by the cosmos – except perhaps very roughly the four quarters of the moon at new moon, half moon and full moon.

What was it in ancient times that allowed the seven-day week, originating in Chaldea, to triumph over the many other ways that then existed of subdividing the month – such as the Sumerians' five-day week, the Romans' eight-day week, the Babylonians' nine-day week or the Egyptians' ten-day week? The answer lies in us human beings. Just as activity and passivity typically alternate during the course of a day, so that, for instance, we are generally least energetic between 1.30 and 2.30 p.m., the human soul likewise resonates from day to day in seven differing moods.

The astonishing thing is that these moods correspond to the qualities and character of the seven classical planets to which each day of the week was assigned in ancient Babylon around 4,000 years ago.

Investigating time today

To investigate time today in a way that takes its lead from the cosmos means both observing the typical character of a day of the week, and also examining the special qualities ascribed to the planet that supposedly governs it. This comparative view will help us understand why, for example, collaborating with others on Tuesday might be particularly challenging.

Knowing more about the qualities of each day of the week, and using these qualities so that we are most effective in what we do, can help us cope better with the demands of modern life and work. In the following chapters we will examine the typical characteristics of each day. This does not of course mean offering fixed prescriptions of any kind but rather giving stimulus to sharpen our perception of time's living, differentiated qualities. Thus each person can set about their own enquiries and reach their own conclusions. This may lead to discoveries such as that school tests and exams done on Friday are less likely to go well than on other days; and that Saturday is the most misjudged and underestimated day.

Rhythm replaces strength.

Sunday: When Does the Week Begin?

Does rest come before or after activity?

Weekend and week beginning

It is clear, isn't it, that the weekend begins on Friday evening and the new week starts again on Monday morning? And yet this division of the week into five working days from Monday to Friday and then two days of weekend leads to problems both of a physical and psychological nature. As the name suggests, we experience the weekend as a well-earned and refreshing end to the week; and the more we enjoy this break, whether by reading, going on excursions or in some other way, the more suddenly and startlingly the new week arrives. This is taxing, and at the same time a good example of the fact that relaxation is often nothing to do with the amount of leisure time we have, but with the organic way in which we structure this time. It is actually part of a healthier way of relating to time that we do not regard Sunday as the last but the first day of the week, as is self-evident in Christian tradition.

Practical consequences

This has practical consequences. If we feel that the new week is beginning on Sunday, we will start looking ahead and thinking about what is coming towards us, and what decisions and plans need to be made. Strengthened by our Sunday rest, we can contemplate the coming week and so allow some of this calm, this inner composure, to flow into the rest of the week.

In his famous statue of David, the sculptor Michelangelo created a work which wonderfully expresses this Sunday mood. Vasari, a biographer of the artist, no doubt sensed something of this when he wrote as follows about the sculpture:

> The diminutive David conquers the giant Goliath with a sling. Why did Michelangelo not express the greatness of this deed by showing in stone how David actually conquers the giant, or even by depicting him standing in triumph over the fallen giant?

Vasari goes on to say that this is because the greatest moment of David's courageous deed does not lie in the battle itself but in the instant preceding the fight, which Michelangelo sculpts: the moment when David, reflecting, takes his decision; and now all that remains is for his will to stream into his limbs. David's clear, grave and confident gaze expresses more of the soul-mood of Sunday than any number of books.

Hectic stress and calm

Many initiatives, actions and activities would be accomplished in a less hectic, stressful and conflict-ridden way if we could form our resolves on a Sunday. Just as rest is needed after activity so that we relax and reflect, it is also needed beforehand, to collect ourselves for the decisions we are making. The conductor who closes his eyes for a moment before raising his baton; the high-jumper who composes himself before starting his run-up, focusing on his bodily feelings: these are instances of little 'Sunday moments.'

We usually connect rest with relaxation, yet Sunday rest can invoke inner expectancy and excitement about things we are contemplating.

To cultivate Sunday, even perhaps to regard it as sacred, means allowing it to become a sun in our lives. Just as the sun inexhaustibly sends warmth and light to the earth, and thus calls forth life and development, so in Sunday, as the first day of the week, we can find a source of confidence and strength for initiatives that unfold on successive weekdays.

Proverbs such as 'start as you mean to go on' or 'in my end is my beginning' do not suggest that events are predetermined, but rather that, like the seed of a plant, a beginning encapsulates the spirit of an undertaking, endowing it with its intrinsic character. In this sense, Sunday is present throughout the week.

Nowhere, nowhere in the world, is there any lack of participating, approving souls. All that is needed is one whose circumstances leave him entirely free to follow his aims with complete resolve.

Goethe, *Clavigo*

Monday: Attending to the New

> Can I hold back my
> response?

Moonrise

The first working day of the week bears the name of the moon for good reason. The moon reveals the quality of devotion in two ways: it is devoted to the sun by reflecting its light and assuming, to our view, the same size as the sun in the heavens; and it is devoted to the earth in that, subduing its own rotation, it couples itself with us and always shows us the same aspect.

In his conversations with Eckermann, Goethe exclaims at one point:

> For heaven's sake, have the courage to attend fully to your impressions, to allow yourself to be delighted, to be instructed also, and to let yourself be kindled and encouraged to accomplish something great ...

In an article on the skills needed in professional biography counselling, I read the following somewhat cutting remark: 'Just sitting there listening is not the same as deep silence.' What the author meant is that far more was needed for openness and empathy in such a therapeutic encounter than a mere invitation to someone to tell their story. It may sound contradictory, but although a therapist has scarcely anything to say in the first few sessions, he must be fully prepared and intimately familiar with the laws of biography. Why? Because at least as much activity can be felt in a person's listening as in their subsequent analysis of another's situation. In seeking to create an open, unprejudiced mood – and this applies to all sorts of other situations – to help the client express himself and find the right words, we have to hold ourselves back. This may sound banal, but it means we must have something specific to hold back. We must have a response to what we hear – our own thoughts, ideas and images – which we are capable of keeping back. The more easily and without obvious effort we manage to hold back our own response, the more richly and trustingly a conversation can develop.

Learning to wonder

In the company of children, particularly, the question continually arises whether we are able to share in their wonder when, for instance, they notice that snails do not pull in their horns in a breeze but do so when you blow on them. Can we keep back our own knowledge in such circumstances?

'Creative spontaneity' makes little sense by itself – it requires some kind of preparation or readiness. This is most noticeable on a Monday. If we experience Sunday as the beginning of the week, using our rest to plan and make decisions about the forthcoming week, Monday can unfold properly. It loses its alarm, and ceases to be the killjoy of the weekend, instead revealing its own intrinsic aspect of bringing something new. On no other day do we encounter so much change as on Monday, and this is why, more than any other day, it requires our openness or even devotion. Yet attending to the new things that Monday brings is only possible when we are looking forward to see whether the ideas we considered on Sunday can be realized on Monday, or whether they were pie in the sky. To find out, we have to absorb reality as far as possible: in other words, allow our own being to grow quiet and at the same time engage it in the greatest activity and attention. This devotion and attention to the outer world is what especially distinguishes the character of Monday.

Openness for Murphy's Law

Naturally, things often go quite differently on Monday than we envisaged the day before; but if we're well enough prepared, our imaginative response will come into play and allow us to cope fruitfully with whatever happens. In this case, Murphy's Law really does apply: if things don't go as planned, be happy that they turned out otherwise.

Attentiveness is in fact not simply a virtue or the result of upbringing, but a state of being, without which we could never be whole. It is, truly, a place where the universe listens in.

Jaques Lusseyran, *The Blind in Society and Blindness: A New Seeing of the World*

Tuesday: A Special Day
for Mistakes

Do we hesitate when
action is called for?

Decisions and plans

Every plan we make, whether to research a country we would like
to visit, or a communal project, or landscaping a garden, takes
place in phases. These separate, characteristic stages can occur
throughout an afternoon or, when time presses, can follow quickly
on each other's heels in a matter of minutes. They come to light
most clearly however over consecutive days separated by sleep, the
most incisive, repeating pause in our lives.

A more or less conscious decision always stands at the outset of
a plan or intention. The more far-seeing and carefully pondered
this is, the more impetus can flow into the plan, and the less likely
doubt and scepticism will be. Once taken, a decision strengthens
our inner sense of certainty, diminishing the impact of external
problems. This is the Sunday mood: making a new beginning
based on calm reflection.

Then comes Monday and resolve encounters actual
circumstances. The freer we are from prejudice and the more

openly we meet and take account of the reality approaching us, the more organically the new can build on the old. The ideas or ideals of Sunday have to connect with what actually happens. On Monday we can often discover how free or otherwise our relationship with our ideas is: are we in love with them or can we stay true to their kernel and dispense with their superfluous shell? The American business author Bill Emmot describes this modern, change-oriented virtue of letting go and keeping hold simultaneously as 'creative destructiveness.'

Actualised will

Actualised will now grows from the dialogue between decisions and experienced reality. The real work begins. In our example of landscape gardening, the decision, say, to build a curved drystone wall may be blocked by the discovery of a concrete post. On Tuesday, therefore, we take a pickaxe and set to work to remove the obstruction, forgetting various other factors in the process. Hindrances that we meet are cleared out of the way, and the desire grows on Tuesday to give increasing shape to those Sunday plans. This day is therefore informed by a particular dynamic and approach.

The threefold dynamic of the planet Mars

Among the planets, we associate the character of the third day of the week with Mars. Just as Tuesday is connected with pressing ahead and the urge for action, so the red planet Mars shows the strongest dynamism in the planetary system. It does so in four ways: in the polar opposites of bright-dark, quick-slow, high-low and calm-tempestuous. For long periods it can only be glimpsed as a glinting, weak orange point of light, but when it approaches more closely to the earth it shines brighter than Jupiter. Its path is unusually compressed, causing it to alternate between a rapid and slow orbit. Only on Mars do sandstorms sometimes shroud the whole planet. Furthermore, the Olympus Mons on Mars is the highest mountain in the solar system, over three times higher than the Himalayas; and accordingly, at ten kilometres deep, Hellas Planitia is the deepest basin to be found on any of the planets.

The interest of mistakes

Encouraging oneself to pursue and realize one's own ideas also involves creating an atmosphere in which mistakes are not only seen as regrettable but are acknowledged and affirmed as necessary aspects of development. Since unconstrained effort belongs above all to Tuesday, it is particularly important on this day to develop a 'culture of mistakes' and regard errors as interesting.

The person who acts is always without conscience. A conscience is possessed only by those who observe.

Johann Wolfgang von Goethe, *Maxims and Reflections*

Wednesday: Interest in What's Around You

What is a centre
without a periphery?

At work, and after

To make a decision in calmness and composure, with openness for and devotion to actual circumstances in the world, and to realize such decisions with dynamism and commitment, are the typical qualities of Sunday, Monday and Tuesday. What now follows the energy and drive that prevailed on Tuesday?

It is something that can no doubt be observed in all forms of collaboration. Here is an example: we are getting a room ready for a festival, with the help of several others. Each person has taken on a particular task such as laying tables, setting up the buffet or arranging flowers. Before long, usually before our own activity has been accomplished, we start to be interested in what the others are doing or have achieved. Whereas previously each person was occupied in their corner, as though wearing imaginary blinkers, we start going over to look at other people's work. Suddenly several people or maybe all are standing together and looking at something in particular.

Concentration and expanding attention

Here a typical rhythm of the human psyche becomes apparent: from concentrated activity, an interest arises in what is happening around us. What have the others accomplished and how does it correspond with what I myself have done? What in my wider surroundings might be of interest and inspiring for my own work? Such questions are typical motivators on a Wednesday. Here we primarily need to pay attention to the work and questions of others. The special potential for this is present on Wednesday, the day of Mercury (as reflected in *mercredi*, the French for Wednesday).

Mercury the relationship creator

A glance from the quality of Wednesday to the characteristics of Mercury can show that deep wisdom, dating back around 4,000 years, underlies the assignment of planets to the days of the week. No other planet engages in such multi-faceted contact with its surroundings as this nearest planet to the sun. Two of Mercury's distinctive characteristics are that it orbits the sun so fast that it overtakes the earth on average every 116 days. Besides this sun orbit, it also spins on its own axis as all planets do. It takes 58 days to complete one such rotation, exactly half of 116, and thus of the time between one earth encounter and the next. Mercury's rotation is coordinated with the earth; but that is not all. Mercury's average distance from the sun, at 55 million kilometres (34 million miles), is 38 per cent of the earth's distance (149 million kilometres/ 93 million miles) and its size, measuring 4,880 kilometres (3,030 miles) in diameter, is likewise 38 per cent of the earth's size (12,740 kilometres/7,916 miles). Thus in both size and distance from the sun, Mercury has the same relationship to the earth.

In fact, 38 per cent is the smaller portion of the Golden Section, a ratio which appears in all growth processes and natural forms. Whereas Mercury's orbit and rotation show harmonious musical time relationships such as 1:2 or 2:3, in its distance and size it reflects the creative proportion of the Golden Section (or the Divine Section as it was called in the Renaissance). However divergent, in fact polar opposite, these temporal and spatial relationships are, they both accord with the fact that the planet has a fruitful, harmonious relationship with its cosmic surroundings.

A further unique quality of this planet underscores the perfection with which Mercury realizes the 'relationship' principle: it places itself freely into its context without tending more in one direction than another. We can take this quite literally: it is the only planet whose axis stands vertically to its orbit without inclining one way or the other. It assumes no prior 'stance' and can therefore enter equally into relationships with all the surrounding planets. Its distinctive quality is to 'enter into relationship with its surroundings' – the mercurial characteristic.

A day for growing

Mercury forms interrelationships with other planets through its spatial orientation and particular type of motion; and attentiveness enables us to do something similar in our own lives. Interest in what's around us, affirming things previously alien, opens up new sources of inspiration. Daily food intake forms the basis for life, for growth. In the same way, external stimulus and the new ideas we gain from it enrich our inner life and encourage spiritual growth.

What is foreign to us has a life that is also foreign, and we cannot fully understand it if we welcome it immediately and gladly like a familiar guest.

Johann Wolfgang von Goethe in a letter to Gottfried Herder

Thursday: Acting with Insight and Overview

How does a correct judgment become just and a just verdict wise?

More than just travel nerves

'Have I thought of everything?' When getting ready to go on vacation, this question can assume dire proportions. 'Did I cancel the papers, are the pets taken care of, do I have my passport, is there anything I've forgotten to tell anyone?' Such pre-travel nerves are often a desperate attempt to make sure we haven't forgotten or overlooked anything. Even when we've reassured ourselves, we can easily feel uncertain again a moment later. Some people are so unsettled by this pre-travel restlessness that it spoils any sense of excitement they were feeling about seeing new countries and cultures. Underlying it, often, is the idea that we might forget something that will later come back to haunt us, since it will be too late then to do anything about it. This is true of other situations too, but is accentuated when we're about to go travelling.

In general this feeling is about a failure in the present that will confront us in the future. No doubt one of the most striking examples of this is the misfortune that occurred during manufacture of the Hubble telescope, which was subsequently sent into orbit around the earth. The wrong piece of equipment was used for polishing the primary mirror, giving rise to an erroneous curvature in the mirror's surface. The mistake only became clear once the telescope was already in orbit. A few years ago this error was largely corrected in a laborious and expensive 'telescope rescue mission' during which a corrective mirror was inserted.

Mistaken judgments and their consequences

How does this relate to the human and social realm? The educationalist Rudolf Grosse gives an example from his time working as a teacher. During a teachers' meeting there was a discussion about a difficult pupil, and the teachers pooled their observations about him: he often didn't do his homework, showed little interest in lessons, gazed into the distance, seldom played with his classmates at break-times, and so forth. The teachers concluded together that the pupil was lazy. 'Bone-idle!' added an indignant teacher. Yet soon after, it was not the pupil who stood condemned but the teachers themselves, due to the school doctor's diagnosis: the boy suffered from a severe digestive disorder. His difficulty passing stools affected his capacity to be mentally alert and involved.

This example aims to show that things we overlook in the social domain can be particularly grave since their consequences often cannot be remedied. These mistaken judgments, the moral condemnation of the boy, might easily have a serious impact on his further development. If we examine our own mistaken decisions and actions, we usually find that the mistake does not arise through false observations or ideas but by forgetting to consider something, by failing to take account of an apparently unimportant yet key aspect. We leave something important out of the picture, and this leads to erroneous actions.

Awareness of the broader context: ecology

The fact that we need an awareness of the broader context when making effective decisions in social situations is true particularly of the quality of Thursday. On Wednesdays we can sense the soul's fundamental mood to be enhanced interest in our surroundings, thus creating for Thursday the necessary conditions for actions that take account of wider effects and ramifications. One might in fact call such actions 'ecological,' because ecology is the study of the relationship between organisms and their surroundings. Thursday is therefore particularly suited for acting in a 'socially ecological' way.

In Babylonian times, Thursday was assigned to the planet Jupiter. What distinguishes this giant amongst the planets? Its diameter, at 142,800 kilometres (88,730 miles), is almost twelve times that of the earth, and it is twelve times slower in its orbit round the sun. The number twelve always indicates a sense of wholeness and completion. Thus the twelve tribes of Israel, the twelve apostles, the twelve knights of the Round Table or the twelve jurors in a court represent the full range of human diversity – or even the whole of humanity. Correspondingly, we perceive reality through the interplay of twelve separate senses, so our processing of and inner engagement with reality is reflected in twelve possible worldviews. Organisation advisors and sociologists therefore often suggest that ideally, to properly evaluate a situation or problem, we should take account of twelve different perspectives.

It is therefore characteristic of this fifth day of the week that we try to develop a full picture in our thinking and act out of this. Put succinctly, the two active, work-accentuating days of

Tuesday and Thursday differ in that action and impetus are
decisive on the Mars day, whereas on the Jupiter day this is
replaced by the possibility of 'wisdom-borne action.'

Men are wise in proportion, not to their
experience, but to their capacity for
experience.

George Bernard Shaw, *A Rebel's Catechism*

Friday: The Importance of What Seems Secondary

How can we make
things beautiful?

Beauty

Our work appears to be finished: we have thought of everything, have considered all circumstances and needs, have even discovered and remedied mistakes. The result is in line with all requirements – and yet one of our colleagues is dissatisfied. He glances over the finished work and makes the surprising suggestion that what has been accomplished still lacks beauty.

It might be a self-built children's play area or a garage for the car. Irrespective of how practical and functional the desired project is meant to be, we wish to add something which is there solely for aesthetic reasons. This represents a desire to liberate ourselves from all necessities and to enhance our work with a special, individual quality drawn from our own creative inspiration. It might be the finishing touch of colour on a wooden shed or a carved shape on the eaves. It might be a pleasing finish to a stone wall in the garden, a vase of flowers on the breakfast table or a special decoration for a cake. Everywhere, in almost every piece of work, a moment arrives when one wishes to place centre stage what is apparently secondary.

Play and imagination

Imagination comes into its own here, allowing us to free ourselves from the diktat of requirements, necessities and conditions. Whereas all practical viewpoints can be derived from prevailing circumstances and originating aims and values – and therefore cannot really be called new – creative enhancement of work gives rise to something new and surprising. It is therefore no wonder that the greatest, childlike enthusiasm is kindled when making something aesthetic. The compliment, 'that's practical' or 'sensible' gives a child much less pleasure, of course, than 'how lovely' or 'that's beautiful'.

Friedrich Schiller's letters on aesthetic education are among the best-known eulogies to imagination: 'For, to say it quite plainly, man only plays where he is, in the full meaning of the word, a human being; and is only fully human where he plays.' In every game we accept self-chosen rules; and in Schiller's view of play, likewise, imagination involves attending to one's own inner feeling of what matches aesthetically, what is harmonious and what not. Without such attentiveness, Schillerian 'play' easily turns to frivolity, and imagination to pure delusion. Beauty then only briefly appears original and soon becomes insipid. It may sound paradoxical, but although there are countless ways of beautifying something, arbitrariness is at the same time the enemy of beauty.

The day of beauty and creativity

Friday is the day of the week when we are particularly inclined to elevate our work into an artistic realm. It is preceded by Thursday, when we draw on insight to place things into a broader context. Thursday can render what we do purposeful, while Friday can make it beautiful. Naturally these characterisations should not be followed too rigidly; but they can help us to recognize the different ways in which our work is supported by the week's rhythmic configuration, and times when we are working counter to the qualities of each day. Friday is less suited to trials and tests, because its creative and artistic mood stimulates us to liberate ourselves from external necessities and conditions.

Babylonian cosmology assigns Friday to the planet Venus. In those times, however, it was named Ishtar, the goddess of love and beauty. Subsequent cultures adopted the essence and qualities of this divine being but gave her different names. Ishtar became Aphrodite in Greece, and eventually Venus in Rome. In Teutonic mythology she was called Freya, which gives us the word Friday; in Romance languages the Roman word was retained (French: *vendredi*).

As Friday is the day of beauty so Venus is the planet of beauty: no other planet emanates light so generously into its surroundings. Our moon, for example, only reflects 7% of the sunlight it receives, absorbing the rest as warmth. Venus reflects back 70%, which is why it starts gleaming before sunset. If we trace the points in the zodiac through which Venus passes when she is in closer proximity to earth, we produce a regular five-pointed star – a kind of 'cosmic blossom.' This picture recurs on the surface of this bright planet, where meteor strikes have not created typical craters as we know them on the moon and Mercury, but flower-like corona structures. These are caused by lava emerging around craters under the influence of Venus's dense atmosphere.

If Sunday can be called God's day, Friday is that of the creative human being.

Do good, and you nurture the divine
plant of humanity; create beauty and you
scatter seeds of the divine.

Friedrich Schiller, *Tabulae Votivae*

Saturday: How Does the Old Enter the World?

How well do we
grow old?

Anyone who has attended the performance of a symphony knows the magnificent moment at the end when the last phrase has been played. With a scarcely noticeable movement of both hands the conductor indicates the conclusion of the final note. The music fades and there is an instant of silence – not fractured by impatient applause only because the musicians and the conductor maintain the tension for a second or two with their intense focus. The violinists still hold bows to strings, the wind players keep their mouthpieces at their mouths, and the conductor's gesture seems frozen in time. Then she lowers her arms, relaxes her shoulders and, in a shared exhalation by audience, musicians and conductor, the applause begins. The piece is over. There are probably few other occasions when we can hear such a full, 'pure' silence as that, just after a concert. In these brief seconds the whole symphony seems to be present in concentrated form. The successive melodies, the rise and fall of different phrases suddenly coalesce in this moment of tableau-like simultaneity. Anyone who has experienced music's timelessness in this way will believe Mozart when he wrote that some of his symphonies were conceived in an instant, and then required laborious days of work to write down all the parts. Even if this far exceeds normal capacities, it nevertheless characterises a typical aspect of the human spirit: the ability to bring different and distant phenomena together in a simultaneous whole.

A time for review

The capacity described above can unfold particularly on Saturday when we look back on the past week. Rather than an analytic review, a strict evaluation of what went well and what didn't, this is more like a kind of 'musical remembering'. What does this mean? New relationships and connections between the week's events appear as we juxtapose them with each other, going beyond known causal connections. We can start to trace deeper-lying causes and motives, in which time also seems to be reversed. For instance, an apparently chance conversation during a train journey on Tuesday may only acquire significance when we find that it bears relation to a task, say, on the following Friday. Thus the reason for the conversation seemed to lie in the future.

On Saturday, the day of Saturn, the week comes to an end. But every period of time, whether a single day or the great cycle of a cultural epoch, shows characteristic phases that start with a new impetus: in a conversation this might be the initial greeting, in the seasonal round it is spring, and for children the new, unwritten school book marking the start of a new school year. In the same way, the end of every developmental cycle is marked by a conclusion: often, by maturation and internalisation.

Innovation and maturity

In recent years the question has often been asked: 'How can something new enter the world?' This question addresses the developmental conditions for innovation. What nurtures new ideas; what favours vision and the necessary corrections to the way we are heading? How do we stay fresh; how does a friendship or working group find the energy to renew itself? Only half the answer is contained in what is new, in development's 'starting conditions'. We must equally ask about the old: how can what already exists ripen so that it gives rise to new seeds? How does the old enter the world?

Whereas getting old at the physical level is something that happens by itself, growing old in soul means growing in maturity, which occurs to a much lesser degree by itself. While Sunday stimulates us to make new decisions and form new outlooks, and thus rejuvenates us, Saturday helps us to grow old by striking a chord from the week's diverse events and experiences, and thus 'harvesting' them.

The creative pause

No doubt because our culture today is so fixated on youth, Saturday tends to be the most overlooked day. Leisure activities and the entertainment industry largely occupy the day when it would make most sense to have a creative pause. Though seeking diversion at the weekend is very understandable, it is important to make space for some peace after the ups and downs of the past week. Perhaps we shy away from this peace, too, because it is always somewhat tinged with sadness – sadness that the process, the specific life of this week is now past. Yet little death processes – when the music falls silent for instance, and equally when we look back at the previous week – are illumined by Goethe's saying about great natural cycles: 'Death is life's greatest invention for having still more life.'

On Saturday, as we revive the events of the week in our memory, they can die away in the best sense of the word and transform into personal capacities. Thereby we become free for new experiences and ideas, and are ready for the new Sunday.

And so the circle of the week closes, the sevenfold wingbeat of the soul.

By far the best experience of men is made up of their remembered failures in dealing with others in the affairs of life.

Samuel Smiles, *Character*

2 | Living with the Rhythms of Time

The Equilibrium of the Present

'Learn to understand what is healthy by looking at what is sick,' stated Rudolf Steiner; and this also applies when we examine our relationship with time. Loss or failure of memory is one of the gravest injuries to our human sense of time or our human existence altogether. Such impairment is widespread of course in milder forms. After concussion, for instance, many people notice that their memory has gaps – that they cannot recall events directly connected with the injury they sustained, as if that time no longer exists and can only, approximately, be experienced through the accounts of others. This little piece of missing past has little impact on our experience of the present, however. It is much more serious when a cerebral haemorrhage deprives someone entirely of the capacity for recall. New impressions or even personal thoughts and feelings can no longer find a purchase in the memory. An impression only lasts in the psyche for the briefest period, scarcely more than twenty seconds, and a veil of forgetfulness shrouds all that is perceived. Such loss of memory means that if you look at the clock, or out of the window to see what the weather's like, or if a friend phones to say he's coming to visit, you forget it all again immediately. You no longer know the origin of the cheerful mood caused by news of a friend's visit, so you assume it is just unfounded cheerfulness. The alarming thing here is that loss of the past also closes off the future: if we can retain nothing in our memory we also lose everything that connects us to the future.

How can this be? Every wish we harbour, every plan we conceive involves an engagement of some kind with the future. We cast a mental fishing line into the future and draw it back into the present. A financial service provider's advert – 'Thinking about the future now' – is well chosen since it applies to far more than merely commercial precautions. The future is always, inevitably, part of our present experience, and every decision, every hope, is directed towards it. In every instant, therefore, seeds germinate towards the future. But in order for them to grow, becoming sources of our actions, we need to protect and preserve them, and this requires memory. It sounds contradictory, but with loss of our memory we can no longer recall our own future.

You might think that this incapacity would liberate us from the freight of memories and visions of the future and allow us to enjoy the present undisturbed, but the opposite is true. Our 'now' is only a rich experience through the presence of past and future. As we digest past experiences and learn to understand new ones, we allow the past to live in the present; and as we set ourselves new aims, we not only think about the future but also create it.

A second apparent contradiction is the following: although it is true that the past and future enrich our present, they can also flood and drown it out. In our experience of the present moment, past and future seem like the two sides of a pair of scales. When they are in balance we feel ourselves 'at home' in the moment. But as soon as one of them preponderates, it becomes hard to really engage in the immediate present. An example from childhood: expectations of Christmas can cause irritation with the intervening period that seems to stand between me and the gifts I'm looking forward to. Worries about the future, or nostalgic memories of a happier past, can overshadow the present. Anyone subject to this feels that the future and past possess a power of their own that can assume excessive proportions in the human soul if we do not continually re-establish a happy medium.

In the zodiac, the sign of the scales stands opposite that of the ram. In almost all illustrations, this ram is depicted in a somewhat unusual stance: it sits facing the circling sun, yet turns its head in the opposite direction. Thus it becomes an image of equilibrium in the temporal realm, a balance between past and future.

By facing the future yet – like the ram – looking back to the past, our present becomes a rich experience.

How Long is the Present?

Many daily gestures such as shaking hands, pointing out a direction, or a welcoming smile, last around three seconds. This corresponds to a single breath, given an average breathing rhythm of twenty breaths per minute. We also clearly sense when we have exceeded this 'moment' because such gestures start to feel forced. Let's look a little closer at this three-second period.

There's a simple test you can try out on yourself, suggested by Ernst Pöppel. Take a metronome and set it at sixty beats per minute. Each second is marked by one beat. With the least experience of Viennese waltzes, we have no difficulty whatever in hearing this uniform beat as $\frac{3}{4}$ time, three beats in a bar – or in other words we inwardly accentuate every third beat. If we take the slowest possible metronome pulse, it is much harder to hear a waltz tune in its monotonous beat, since 3×1.5 seconds until the next accented note far exceeds the three-second duration of a breath. Only within this span of time can we easily gain the sense of a bar, a group of beats. Without having to remember, we know when an accent is due again, as all three beats, despite sounding in succession, are 'compiled' in our mind into a present. The same is true in conversation. Even a longer sentence is understood when spoken, without us having to actively recall the beginning of the sentence. The succession of words, and even of syllables, is something we integrate into a non-successive whole. This does become harder, however, if pauses, especially at inappropriate points, exceed the span of three seconds.

'Now' is, in human experience, not just a skin-thin boundary between future and past, between 'in a moment' and 'just now,' but has a temporal duration that only we ourselves produce. The sand-clock as an image of passing time is only therefore appropriate as a physical measure. Soul time does not flow continuously in the same way, but runs in single waves, each of which endures for three seconds as a small parcel of time. The 'now' feeling arises as we merge together successive events into temporal units.

This is why watching and listening to plays, dance or music is a good way to school our sense of time. In these 'time arts', present moments are created that are longer than the breath. Whether as melodies stretching their arc over many bars, the entirety of a dancer's sequence of movement or an actor's soliloquy, the audience can in response develop more elevated present moments. A drama or a symphony can ultimately transform in the soul to a single, whole presence.

In these higher and more encompassing forms of 'now', the pervasive diktat of cause and effect, of 'must ... because' is suppressed. The laws of physics and causality pale, and suddenly we can have an artistic experience of a conceivable, opposite realm: a later cause can actually give rise to an earlier event. For instance, two people do not live together because they once got to know each other and came to love one another, but the reverse: they had to meet in order to live together as they do.

Our physically enlightened reason is wrong to prevent us looking at fate as though the cause might lie temporally after the effect. Denying this reality of the reversed stream of time will hinder us from experiencing the magic and wisdom which, despite all chance occurrence and freedom, lives in human destiny.

There are many people who have been revived after a severe accident and who speak of having a near-death experience. Their accounts are unanimous in describing their life passing before them in a tableau, a kind of simultaneous panorama. Neurologists often seek to explain away this phenomenon by saying that all our memories are released at once in the dramatic state of oxygen deprivation. Yet this higher simultaneity is nevertheless supersensible in nature, in the same way as we partake a little of the supersensible realm in the enhanced experience of immediacy we can gain from a symphony, book or play.

Watching and listening to plays, dance or music is a good way to school our sense of time. In these 'time arts' present moments are created that are longer than the breath.

Speed and Heart Rhythm

For many weeks one summer, the sculptor Mirella Faldey moved into an open wooden shed close to my office and set up her workshop there, creating four open-air sculptures. She had sought out the hardest stone, granite, as her material. To work it you need specially toughened steel chisels which emit a ringing tone when they strike the stone: 'ding, ding, ding'. After four weeks I had got used to this regular, bright sound so that it more or less merged into the other noises in the vicinity. But then came a surprise. One day, suddenly, the chisel strokes disturbed me again, but in a much more penetrating way than on the first day, making me unusually nervous and unfocused. Looking through the window to the shed I saw that a male sculptor had replaced Mirella and was tackling the stone with special energy. Although his chisel strokes were no louder or less regular, I could barely stand the noise. Why? Looking at the clock on the wall gave me the explanation. The assistant sculptor was using his hammer and chisel 170 to 180 times a minute, or in other words three times a second. This was a good deal faster than his colleague, who worked at a calmer tempo of about 140 to 150 strokes a minute. The difference of thirty strokes more per minute seems small, but is decisive. Whereas the slower rhythm still resonates in a somewhat musical fashion, the faster rhythm has an insistent, mechanical quality.

The point at which a rhythm loses its sense of naturalness and ease is due to an elemental human rhythm. Besides the heartbeat rhythm of around 75 beats per minute, there is another heart rhythm, roughly twice as quick, which is caused by the response of blood vessels to the heartbeat, their elastic absorption of the pulse. After every pulse beat the pressure wave spreads from the heart through the arteries into the whole body. The finely branching system of blood vessels reflects the pulse beat back towards the heart and allows the blood to flow back a little. Since the heart chambers are now closed, the blood cannot flow back into them but instead surges from the closed heart back to the periphery of the circulation system. Thus every original heartbeat is followed by the echo of a second, weaker pulse, visible in an ECG as a smaller wave. The blood not only resonates in the pulse rhythm of 75 beats per minute, but also in the double frequency of around 150 beats. This 'overtone' of the blood is known as basic arterial vibration and is the quickest fluctuating rhythm in the body.

But as well as setting a temporal measure in the blood's circulation, basic arterial vibration acts as a temporal sound-reflector in the soul. This is why hardly anyone is likely to wish to, or be able to write a personal letter at the same speed as a phone number noted down in passing. To put our personal stamp on handwriting we need to write no faster than this fundamental resonance allows. Our gait also unfolds within this range. At more than 150 steps per minute, movement becomes mechanical and hectic.

While the heart only beats in the chest in spatial terms, we rediscover its temporal influence in walking, writing, hand-shaking, craft work and music-making – as long as it is not ousted from these activities by excessive speed.

How Long Does a Moment Last?

If we assume a moment to be, literally, the blink of an eye, the answer to this question is six seconds. That's the period of time in which eyelids blink once on average. They slide down rapidly over the eyes, re-distribute the tear fluid, clean the eye's surface and give a brief respite from seeing, a swift moment of rest. Yet this fraction of a second pause in sight is decisive because it rhythmically structures vision. We look out into the world with open eyes, departing from ourselves because our sight takes us out into the phenomena around us, but the eye-blink* brings us briefly back to ourselves. It helps us, for instance, to regain a little detachment after complete involvement and absorption, so that we don't entirely lose our capacity for judgment in self-forgetful gazing.

The rhythm of the blinking eye is not only different in each of us but also varies depending on the work we are doing and our inner state of mind. It can speed up to be as quick as the breathing rhythm, with the lids closing every three seconds, or it can slow down to just a few times per minute. The latter is likely in a reverent or contemplative mood, whereas in a heated discussion or when we feel uncertain, the eyes can blink frantically.

We can do a little test on ourselves: if we look at nature or at a

---

* In German the word Augenblick (literally 'eye blink') means 'moment'.

painting and blink frequently as we do so, we find that a certain span of uninterrupted vision is needed for us to work through an impression in our feelings. Too frequent blinking prevents us fully absorbing or participating in something new.

The eyes are always involved in absorbing information, are intellectually challenged, whether we're driving, working at a computer or reading. It is advisable to balance this by allowing the eyes to dream, too, from time to time – in other words to let colours and forms pervade them. The eyes have to grow calm in order to enjoy the yellow of a blossoming canola field or the colour nuances of a sunset. This seems to happen by itself the moment we turn to such an impression in a focused but open way.

The psyche configures the body's temporal sequences to suit its activity. As the physician and cosmologist Walter Bühler puts it, the body becomes the instrument of the soul. The blinking eye is a kind of soul breathing. It grows calmer in a reverent mood, and its frequency increases with intellectual activity, for instance when we're trying to find our way in an unknown area, or when we're looking for something we've lost.

In Bertolt Brecht's play *Galileo*, the dramatist has the Italian astronomer explain to his housekeeper how the movements of the planets should be seen heliocentrically. To demonstrate this he sits her on a swivel chair and starts spinning it. When he asks her what she saw she gazes at him with incomprehension, and he flares up at her, saying, 'You shouldn't just stare, you should look!' He means that we can only understand what we see by briefly turning away after observing something: then the immediate impression recedes and space is created for thinking.

After blinking we look out again afresh into the world. The six-second blinking rhythm amounts to 5,000 moments of fresh vision every day. The blink of an eye is therefore sleep's little brother, just as sleep itself is known as the little brother of death. As we blink, the ongoing process of sight 'dies,' but that is precisely what makes it human. Just as we rediscover ourselves anew after a night's rest, and can then look on the world again with new eyes, so every blink renews our eyesight.

One of the most charming ways of greeting someone is with a wink, as the renewal and refreshing of sight is not for our own sake but for another.

After blinking we look out again afresh into the world.

The New Day Begins
in the Evening

Transitions that involve passing from one state to another, from one form of life to a different one, are the most spiritually active moments. We often only notice this later, since only in the safe haven of the new phase of life does it become clear how lively and active the period of change was. What may have been experienced at the time as uncertainty or crisis is later revealed as a phase of special intensity.

Such transitions may be major stages of development, such as that from youth to adulthood, from training to the practice of a profession and family life, or in advancing towards old age when various personal and societal responsibilities are gradually taken from our shoulders. But besides these periods of major change, our biographical path seems to hold its breath and intensify at other moments. On a smaller scale, too, human time is configured in differentiated ways, and made rhythmic by intervals and pauses. This happens most regularly in the alternation between day and night, between waking and sleeping. Comparable to the bar lines of a piece of music, this rhythm forms a fundamental oscillation in human life. Here too, the transition moments between these polar states have a particularly intense quality.

In the same way that sunrise and sunset offer the greatest wealth of colours in the day, this transitional, gloaming period offers us special possibilities. Nature cools and comes to rest at twilight, and the same is true of us. Our pulse and breathing grow slower from late afternoon onwards, blood pressure sinks, and our speed of reaction slows. A decisive change also occurs in the body's warmth system: while temperature sinks at the forehead and internally, it rises in the hands and feet. Just as the earth rays out the heat of the day into the cosmos, our own body heat likewise turns towards the periphery. We therefore need to keep ourselves warmer than in the morning.

But as well as our need of warmth and sensitivity to cold, our awareness of noise also rises in the evening. As light fades, not only our warmth expands but also our hearing capacity. Sounds from afar now reach the ears. Every creak in the house or noise from the street is perceived more attentively. These phenomena accentuate our experience that in the evening the soul releases itself from close focus, from the day's preoccupations, to widen and expand. This is why we are especially receptive to 'food for the soul and spirit' during evening hours. This might involve a concert or trip to the theatre, inviting guests for a meal, a short walk or reading an interesting book. The evening is the time when we are most able to discover new questions or absorb different perspectives.

In the evening, the day's events can once more be invoked. While an exciting crime-thriller film may 'rivet' us, cultural events in the evening, a concert or theatrical performance, can support this expansion of the soul and can often act as catalysts to help us draw some reflective profit from the day's events. However paradoxical it sounds, this occurs primarily through questions. The best harvest of the day is if something we experienced appears to us as a riddle in the evening and loses a little of its apparently self-evident quality; for questions about the way things are always mark the start of new development. This applies not only to doubt and scepticism, which question things in a disengaged or dismissive way, but more to questions with some aim, some personal connection – which always, whether we know it or not, inherently contain something of their answer.

From Discoverer to Philosopher in the Cycle of the Year

In the seventies, the German railways launched investigations into the frequency of errors made by engine drivers. Were there particular times when signals were more often overlooked and top speeds exceeded? The company found the well-known phenomenon that most mistakes occur in the early afternoon, between two and three p.m. At this time of day there is the greatest risk of briefly falling asleep. A full stomach after lunch is not the prime reason for this, since people showed symptoms of exhaustion even if they ate no lunch. The railway investigators also discovered that, following the introduction of a day off on Saturday, trains were more often delayed on Mondays. After two work-free days, the organism has to adjust again to work, and this, according to rhythm researcher Gunther Hildebrandt, was what led to the operational problems.

But besides these two times of increased tendency for errors and exhaustion – the 'low-point' in the afternoon, and on Mondays – the investigators found a fluctuating performance graph through the year: in December, January and February a third more errors occurred compared to the same length of time in spring (April/May) and the start of the autumn (October), when there were least.

Central heating, imported vegetables and fruit from southern lands in the winter, and air-conditioning in the summer, detach us from our connection with the seasons. The enquiries of rhythm research can help us to take proper account of the seasonal cycle as a series of temporal transitions reflected not only in the leaves on the trees and the length of the days but also in our mental and physical state. A few examples can illustrate this. During the winter months our reaction times are slowest. We respond relatively dully to external stimulus and there is also a widespread tendency to resignation and despondency. What do the findings of rhythm research tell us here? They show not only that physical performance is greater in summer but also our capacity to gain physical and mental stimulus and inspiration from the external world. In the winter, the external world's influence grows weaker. Our sensory activity fades in autumn as nature comes to rest and the life forces in perennial plants and trees withdraw. With the dark time of year, our attention turns from our surroundings back to ourselves. Whereas we immerse our souls in summer in nature's diversity, in growth, blossoming and the humming of bees, and in our fellow human beings with whom we converse, in winter we withdraw more, instead, into the cosmos of our own souls. In summer we tend more to be discoverers, in winter philosophers and artists. The fact that rhythm research comes up with mostly negative findings for the winter, as stated above, is due to the difficulty in measuring this seasonally-governed philosophizing and artistic sensibility. That reaction times are slower in winter should not lead us to conclude that we are sleepier. In fact the opposite is the case: in winter we are wide awake because we are fully in ourselves.

In winter's peace and interiority we develop our steadfastness. By drawing on the potential of this season for self-reflection, and pondering on our relationship with worlds above us, we create the foundation for fully engaging again in the dynamic activity of spring and summer.

On the arena of the human soul, therefore, summer and winter join hands. In nature, the seeds of autumn and winter start pushing into growth in spring, forming relationships with their surroundings through roots and leaves. In the same way we lay soul-spiritual seeds in winter, which can become reality in spring and summer.

Ex-Beatle George Harrison, who died in November 2001, coined the following phrase: 'Everything else can wait but the search for God cannot.' We can give this phrase special importance in the dark months of winter.

Everything else can wait but the search for God cannot.

The Will's Small and Great Rhythm

'The greater is reflected in the smaller' is a prime law of living things. For instance, as well as seeing human soul activity in relation to the human body – so that the head equates with thinking, the chest with feeling and the limbs with will – we can go further and seek all three realms of the psyche in the head itself. Then we find that the forehead corresponds to thinking; nose, eyes and mouth, as the centre of the head, express feeling; and the chin and jaw as the head's only moveable part correspond to will. The hand is another example. We grasp, point and hold things with it, and so it is an instrument of the will; and yet the fingertips possess a perceptive capacity typical of our head and thinking. The expressive quality of the palm and back of the hand in gestures shows that this region is dominated by feeling. That's why we find it strange if, when shaking hands with someone, they only take hold of the front part of our hand, the fingers. In greeting someone the hand surfaces have to meet so that we briefly perceive each other there. The thumb is really the only inherently will-oriented part of the hand.

The whole is reflected in the part: this is a characteristic of every living thing, not just in relation to its spatial but also its temporal form. Thus in nature, the great structures of time can be found in small ones. Differing measures of time are related to each other in the same way as the course of the year and the day. What we experience as freshness and cheerfulness in the morning corresponds to spring in the seasons' round. Spring is the year's morning, and therefore the Easter festival has a morning quality while Christmas is a night-time festival. In the same way we can sense a connection between summer and midday, and the afternoon and autumn.

But there is a less familiar relationship between two human rhythms: the moon takes exactly one hour to move the distance of its own diameter across the backdrop of the stars. The same time is needed for something strange or new to become fully a part of us. Whatever we preoccupy ourselves with for an hour, whether a thought, an artwork or a plant can, to the extent we manage this, take such deep root that it makes us into a different person. This is the rhythm of the human will. If we take an hour to think about someone else, our relationship with him can acquire lasting depth and honesty.

We can ask whether this will rhythm of an hour (or $1/24$ of a day) corresponds to a comparable period at the level of a whole human life. The calculation produces this result: 72 years divided by 24 = 3. What the hour represents in the day corresponds therefore to three years in a life. If we think of the first three years of life, of early childhood development, we see that everything which we need later in life is formed then: walking, speaking, thinking, use of the senses and, as the crowning quality, the ability to experience oneself as an individual, as 'I'. This breathtakingly quick development occurs – and this is the mysterious thing – without parents having much influence on when and how each step takes place. Neither the will of the child nor the will of the parents makes the child tirelessly practise grasping, crawling or standing. This is cosmic will at work, a gift of the heavens, a festival that lasts three years. It mirrors the archetypal span of time from the Jordan baptism to Good Friday three years later.

3 | The Quality of Numbers 1 to 31

Introduction

God made all the numbers; everything else is human labour.

Leopold Kronecker, in a talk in 1886 given at the Berlin meeting of the Society of German Scientists and Physicians, and published in the annual report of the German Association of Mathematicians in 1893.

One sun, two parents, three meals a day, four seasons and five fingers: children soon discover that most things in life exist in a particular number – which is more than just a sum total but actually expresses something, some quality of the thing in question. Thus the one sun shines on each of us, assuring us that we live in one world, a whole, a unity in which it shines. Two people lead us into life and two is visible everywhere: in above and below, good and bad, forwards and backwards, sleeping and waking, light and darkness, and even freedom and compulsion.

From the seven dwarves and the same number of notes in a scale, or the 23 chromosomes, through the 32 types of crystal in mineralogy to the 153 fishes which the disciples in the New Testament haul ashore from the Sea of Galilee, numbers in nature, culture and religion tell us something about how things are constituted, about the inner workings of the world. Yet ever since numbers came to designate a time of day, a distance in miles, or have been found on bank statements and price labels,

Plato's phrase about the world conceived in the language of mathematics has acquired a quite different resonance than intended by the Greek philosopher. Plato's outlook has been superseded by a dictate of modern science formulated by Galileo Galilei: 'Science is what can be calculated – and what cannot must be made so.'

From number to wonder

Being able to measure and enumerate answers the question of *how much*, and helps us order the world so that we grasp it in a system; can plan and build within it. But this does not tell us *why*. It scarcely helps us to understand the world and the meaning concealed in it. The Swiss philosopher Karl Barth distinguished two forms of knowledge.

Knowing that aluminium conducts heat and electricity, understanding ATP metabolism within a cell, or being able to place the stability of metals in an ordered sequence, provides a basis for manufacturing machines and medicines. It is 'access knowledge' because it gives us access to the world and its immeasurable wealth of things and creatures. This knowledge is the tool by means of which we have, more than in any other age, both pursued and misunderstood the Old Testament phrase of 'dominion over the earth'.

There is another type of knowledge however, which also enumerates: not in order to amass number but to gain insight. Termed 'orientation knowledge' by Barth, this knowledge gives relationship rather than power, participation rather than dominance; and repeatedly, the feeling which is the beginning of all philosophy: wonder. We are struck by wonder that there are seven oceans, seven colours, tones, and orifices in the head; and from this recurring number phenomenon an image forms so that step by step the inner characteristics of a number are revealed. Being, by its very nature, cannot be proven but must be felt and sensed. In searching for clues, mathematics, the sciences, culture and religion come into conversation with each other. It is part of the mystery of numbers that their intrinsic nature becomes apparent as a bridge across mathematical and cultural divides. Thus for instance, the perfection of six can be found both in Babylonian cosmology and religion, but also in the domain of arithmetic.

The calendar limit

When I embarked on these little monographs on the numbers, my scope was initially limited to 12 or at most 17. While I knew that we have 24 ribs, and that 28 is a perfect number, at the same time I imagined that after 20 it would scarcely be possible to stake a claim on the personalities of the numbers. I was all the more taken aback, therefore, when I became aware that even a number like 29 or 31 has a distinctive character which sets it apart from all others. Yet where do we stop? Naturally there are also greater, interesting numbers which soar above the landscape of numbers like mountains – such as 33, the number of the sun and of Christ's life, or the number 257, a prime number* polygon that can be constructed as geometric solid, or Plato's befriended numbers 220 and 284. In this book, the human being or the calendar has set the limit: our birth date gives each of us a particular relationship with one of the numbers up to 31.

The 31 accounts, of which 1 to 24 previously appeared in the journal *a tempo*, are excursions into the realm of number, their typical locus in the natural kingdom and in culture, which aim to help us understand anew Plato's exclamation that 'the gods do geometry'.

---

* Prime numbers can only be divided by themselves, and not by any other number. In other words they have no factors.

The spiritual character of numbers

While a child must be taught to write and shown the letters, counting is a little different. Children intuitively start to walk in a rhythm over the joins between pavement slabs and to whisper numbers as they do so. Later one finds that there is hardly any area without a relationship to numbers. Numbers order nature: from the arrangement of petals in the rose and lily through to the planets, where five belongs to Venus since, with the earth, it inscribes a pentagram in the sky, and twelve to Jupiter because this giant among planets takes twelve years to make its way through the zodiac, and is twelve times as big as the earth. Whereas natural constants such as the Euler number (e) 2.71828... and the number of the golden section (ϕ, phi) 0.61803... as well as π (pi) 3.14159... are irrational – in other words have sequences that are apparently infinite, the natural numbers we are concerned with here have an unmatched simplicity. This simplicity reveals numbers' high spiritual character. Rudolf Steiner showed yet another aspect of numbers when highlighting their spiritual value. He reminded us that we can do anything we like with them: although they serve to describe the highest, divine nature – for instance the Christian trinity, or the three Indian gods Vishnu, Shiva and Brahma – they can also be used for the most profane or even destructive purpose. According to Steiner it is precisely this selfless nature of numbers which elevates them – or rather the spirit that casts them, as it were, as shadows – into high spiritual spheres.

This book is a doubtless incomplete attempt to trace these shadows cast from the eternal realm so as to interpret them in a way that Plato would have recognised. The provisional nature of the accounts can serve as an invitation to readers to extend them themselves, by seeking out number phenomena in nature and culture. I wish every reader some of the inspiration and wonder that numbers are able to kindle when we acknowledge, with Pythagoras, that they 'are the essence of all things'.

1

The Number of the Whole

The happy man is the man who does not suffer from either of these failures of unity, whose personality is neither divided against itself nor pitted against the world.

Bertrand Russell, *The Conquest of Happiness*

We begin with the most difficult number, the number one. The medieval mystic Agrippa von Nettesheim wrote that 'Unity penetrates every number and itself always remains the same'; and the mathematician Köbel wrote in 1537 that one was not a number at all, but rather the endower, the beginning, the foundation of all other numbers. We come to such an idea if instead of seeing the numbers additively, we view them as divisions of the one unity. Then one is the whole from which the other numbers grow by division into two and beyond. Although one is the smallest whole number in mathematical terms, it embodies the greatest dimension of what we might call 'unique'. As people we are not just *one* personality with *one* history, but we see the world as a *unity*. This idea of the unity of the whole world sounds elementary and yet is one of the great ongoing processes of human insight. It takes its departure from the idea that there is only one God, one Creator, rather than a host of divinities such as we find in all pagan religions.

This is connected with a profound change in human consciousness, for only when we believe in one God is it possible also to speak of an enclosed world and a distinct personal identity. The pharaoh Akhenaton was the first to replace the countless figures of gods and spirits with the sole sun god Aton, thus elevating *one* to its throne. Judaism, and subsequently Christianity and Islam, likewise spoke of one Creator, enabling the indivisible, all-encompassing element to inform the human spirit.

Philosophically, this question as to whether the world is one or a larger number, has preoccupied almost all thinkers. In ad 250, Plotinus wrote, 'Every multiplicity is a multiplicity of unities, and therefore is predicated on unity.' Eight hundred years before this, the Greek philosopher Parmenides had already sought to grasp the nature of one in a didactic poem, by comparing the unity of the world with the form of a sphere.

The image of a sphere as one, is an image which cosmology draws on when speaking of the curvature of space, which is finite but boundless. The cosmos must be finite for if it were infinite there would be an infinite number of stars with infinite light, and at night the sky would be glittering bright. Nevertheless, due to the curvature of space there is no limit to the cosmos – in the same way as the surface of a sphere does not end. Just as the square is used as an image of the number four, so the circle or sphere is the image for the greatest number, one, which according to Pythagoras is the only number that is simultaneously masculine and feminine.

At the age of one, we can stand 'on our own'; there is one sun, which sustains life, and one earth. There is a first breath at birth, and a first day at school. No number has such an aura as this, standing at the beginning of everything and therefore less easy to grasp and understand than all other numbers until – usually after a long quest – we become able to grasp our own identity, the quality that distinguishes our own 'I'. I suspect that this is the key to the number which is at the same time the biggest and smallest of all.

2

The Number Between
Doubt and Suspense

Question: Look at a stick –
One end is yin, the other yang.
Which is more important?

Answer: the stick is more important!

Lao Tzu

'I have two friends, and can imagine marrying either. How do I make the right decision?'

This was a question which a female student put to Georg Glöckler, a lecturer in Waldorf education. The answer he gave greatly surprised her, 'Flip a coin. If you really don't know, then let heads or tails decide it.'

The student replied: 'But what if the coin tells me to marry A, yet at that moment I find my heart belongs to B?'

He had been waiting for this rejoinder, and smiled as he said, 'Then give A his marching orders, and be pleased you've found the true answer.'

A good proposal: the coin helped to discover the decision that the heart had long since made, whereas the intellect was still caught in the back-and-forth of reason. 'How do we progress from two to one? This is one of the most important biographical questions,' added Georg Glöckler.

An unresolved stalemate situation can tell us more than almost anything else about the nature of the number two. Two is doubt, dispute, discord, division, hybridism; two is, as the poet Friedrich Rückert writes, 'the twin fruit on the branch, both sweet and bitter'. In the form of *duo* and *dubio* in Latin, and *doubt* in English, it embodies a double and dubious condition. In number mysticism it is seen as the number of abandonment or loneliness, of evil – for it signifies standing in opposition to the one, the divine. And here lies its contradiction also: one encompasses everything after all, so how can one be outside or in opposition to the all-encompassing? Only by standing outside the divine and the world itself. Another poem by Rückert states: 'I have become lost to the world', which Gustav Mahler turned into one of his most beautiful songs.

Two is the only even prime number – and in this peculiarity lies much of the secret and contradictoriness of this number. Just as one comes into being as the point, so the line represents two: it connects two points and divides two surfaces. Novalis writes that touch is 'simultaneously separation and uniting'. The division of human beings into two sexes is at the same time the source of all capacity for connection. Two stands between one and three. While this phrase seems ridiculously banal, study of its implications could fill whole libraries. Two is the number of opposites: day and night, good and bad, Cain and Abel, yin and yang.

> In Taoism, yin originally refers to the colder north side of a mountain and the shady south bank of a river, while yang denotes a mountain's warmer southern slopes and a sunny, northern riverbank. In Chinese philosophy, this gave rise to principles that permeate the whole of existence. Corresponding to the flourishing of nature in the warm season, yang stood for everything active, engendering, enlivening, creative, shimmering and outward. Yin, meanwhile, corresponding to wintry qualities, stood for all that is passive, hidden, non-luminous and inward. The two enhance and determine each other, and find their state of equilibrium, their cosmic marriage as it were, in the summer and winter equinoxes.

A hundred years ago, the quantum physicist Nils Bohr discovered how deeply two is inscribed in nature. In relation to the phenomena of light and electricity he showed that nature always possesses two faces, one of whose contradictory countenances is revealed depending on the type of experiment used. Bohr compared this complementarity within matter with the opposition between love and justice: pure love is always unjust, and pure justice always loveless. They are opposites; and yet our humanity only unfolds fully in their interplay.

3

The Queen of Numbers

Three has a structuring and organising quality.

Rudolf Steiner, *The Cycle of the Year as Breathing Process of the Earth*, lecture of April 2, 1923

Heads or tails, left or right, yes or no? Where two roads diverge there is no freedom. This dichotomy is beautifully described in Robert Frost's poem, 'The Road Not Taken'. For each choice in favour of one must inevitably be a refusal of the other. We can see this in the dire saying, 'Who is not for us is against us'. Only when the famous 'third way' is available as a mediating choice, does a free interplay begin in which, for instance the cosmos of colours arises from black and white. According to Honoré de Balzac, three, along with seven, is the spiritual number *par excellence*; and, as Aristotle says, it is the only number that has beginning, middle and end – and only one of each. This is no doubt why three is the queen among numbers, and occupies a central position in all religions.

Sumeria, 7,000 years ago, had Anu, Enlil and Ea (heaven, air and earth); ancient India had the creator Brahma, the destroyer Shiva and the sustainer Vishnu; and Egypt had Isis, Osiris and their son and redeemer Horus.

The three planes of space, matter as solid, liquid and gaseous, and time as a triad of past, present and future, go hand-in-hand with what Plato describes as the ideals of beauty, truth and goodness, and Paul as faith, hope and love.

In fairy tales three wishes are given, three sons set out on adventures, and three golden hairs must be pulled from a giant's head. Later on, three musketeers seek freedom, and seven times three astronauts fly to the moon.

'The Tao engenders unity, unity engenders the two, the two engenders the three, and the three engenders all things,' says Lao Tzu in a country in which wisdom rests on the three pillars of Confucianism, Buddhism and Taoism. This means that three is the bridge between the one and the many. A sentence is composed of three prime constituents, subject, object and verb. What in Caesar's time passed into history as the Triumvirate – the political friendship of three men – is still current as the archetype of a successful team, and can be found in its most natural form in mother, father and child.

In three we celebrate a resurrection, and this is why the number pervades all strata of Christianity. Three shepherds and three kings bear witness to the beginning of the life of Jesus, three crosses stand at its end, and three days are required for the Resurrection. 'I am the way, the truth and the life.' In this inner trinity is reflected the grandeur of Father, Son and Holy Spirit. And Christianity itself celebrates three great festivals each year, at Christmas, Easter and Pentacost.

One can enumerate countless places in nature and culture that reveal a threefold quality, as triad, trinity or tripartite whole. We can repeatedly discover that the step from two to three opens a cosmos of possibilities. Physics calls this diversity the three-body problem, and refers here to the impossibility of calculating the interplay of three physical forces. We ourselves are the most magnificent locus of this interplay of three forces: what appears in visible form as limbs, trunk and head, and works in three different ways as muscles, organs and nervous system, embodies in the human personality the presence of an individual spirit in will, feeling and thinking.

The Number of the Earth

And after these things I saw four angels standing on the four corners of the earth, holding the four winds of the earth.

Book of Revelation 7:1

Four is the first number that possesses factors. In other words, it can be built up from other numbers – in three ways: adding, $2 + 2 = 4$; multiplying, $2 \times 2 = 4$; and squaring $2^2 = 4$.

If two is the number of opposites, four is the number of the dynamic of these opposites. And indeed, four can be found in all forms where the interplay of opposites is at home: in the earthly realm, the sensory world. The four points of the compass order space, and matter can have four forms if we add to the solid, liquid and gaseous states what physics calls plasma (under extreme conditions, as in the interior of a star). The name Adam, the 'first inhabitant of space', bears in Greek the four letters of the points of the compass, *anatole, dusis, arkto* and *mesembria*.

Since Einstein we have assigned four dimensions to space-time. Animals walk on all fours, and the car drives on four wheels. Physics knows that four fundamental forces maintain the world's coherence: in the larger world these are gravity and magnetism, and in the miniscule world of the atom, we have the weak interaction which keeps molecules together, and the strong interaction which joins particles together in the atom. Likewise, there are four forces of the psyche in the temperaments. Based on the ancient world's view of the four elements, the soul can similarly be seen as solid (melancholic), liquid (phlegmatic), airy (sanguine) or fiery (choleric).

For the past hundred years, physics has been seeking a unified account of the four fundamental forces, and dreams of making these four physical forces into one with a universal theory. This is likely to remain a dream, for it is inherent in the world of objects, plants, animals and human beings that two, three and above all four have replaced the original one. In the Egyptian myth, the dismembering of the god Osiris leads to this loss of unity; and it is scarcely surprising that almost all world religions have a fourfold response to the all-embracing one: there are four gospels in the Bible, and also four sacred books in Islam – Torah, Psalms, Gospel and Qur'an. India likewise has the four Vedas.

The most magnificent expression of four can be found in the Egyptian pyramids. Much has been written about their geometry and proportions, and new discoveries are continually being made. Pi (π), the ratio of the circumference to its diameter, was found as relationship between height and side length; harmonious triangles and the pentagon appear in the side planes, but the essential quality of the pyramid is summed up by saying, simply, that one gives rise to four. At the top, near to heaven, the golden tip of the pyramid is enthroned, and in sunshine becomes a gleaming point of light. Below, its four-square base gives it firm purchase on the earth. In the most primordial form we have embodied here the path that everything ideal takes from a unity to the fourfold nature of the material world. The square itself embodies four in a similarly powerful image. No form radiates such sovereignty and stability as the uniform square or its brother the cube.

Two thousand five hundred years after the Egyptian pyramids a new image acquires similar significance, imbued also by four and one. This is the cross, whose message is now reversed, for its four arms run to a central point, a crossing point, which in the Celtic cross was accentuated by a sun wheel, and in esoteric Christianity by a circle of roses.

5

The Human Number

Mephistopheles:
    I fear I'll have to be so bold
    as to tell you: a little obstacle
    prevents me leaving here – the pentagram
    at your threshold ...

Faust:
    The pentagram is hindering you is it,
    you son of hell? If that's what's kept you here
    how did you ever find your way inside?
    How was such a spirit tricked like that?

Mephistopheles:
    Just take a look: it is not drawn quite right ...

Johann Wolfgang von Goethe, *Faust, Part I*

This number has its celebration in the months of April and May – first in cherry, pear and plum trees, then on rose bushes everywhere. Despite all explanations by specialists in botany and plant physiology, it remains a miracle that the number five blossoms forth from brown branches and green shoots in white, pink, yellow and red blossoms.

Naturally there are also blossoms with three petals, such as the tulip and lily; or four, like the buttercup. But the rose, the Rosaceae family, remains the archetype of blossom.

> If you cut open an apple or a strawberry, the five-pointed core or seed pith offers a reminder of the blossoming five of spring.

To understand five, one must understand the blossom. Whereas growth and metabolism are focused in the root and foliage, the plant intensifies in the blossom to something approaching soul expression, through colour, light and fragrance. In the blossom the plant's typical character of growing ever onward in leaf after leaf, branch after branch, comes to a standstill. Through this 'stasis' the blossom as it were grows beyond itself.

The geometric image of five, the pentagram, has a similar quality. In contrast to the triangle and square, a star arises in the sequence of lines in the pentagram – a special one, in which all line segments mutually bisect each other in the golden section. This is a division in a specific ratio in which the shorter part of a line relates to the longer as the longer to the whole. In the Renaissance, this proportion was rightly called the *sectio divina*, or 'divine division', because, alongside the close relationship between two sections, it also

contains a mathematical secret: no matter into how many smaller parts you divide one of the sections, they can never fill the other section without leaving a remainder. In mathematical terms this means that the golden section is a division in which the two sections are least concordant with each other, are as alien to each other as it is possible to be; and yet – and this is the magic of the golden section – give rise to a higher affinity. This is why nature adheres everywhere to this proportion – for instance in the length of tree branches, the shape of leaves or the sequence of finger joints – since, more than any other ratio, it embodies the principle of bringing forth something new without loss of unity.

> For the ancient Greeks, five, as the *Quintia essentia* (via Latin) was elevated above the four of the elements and symbolised the cosmos, the spiritual realm. During the Renaissance the cosmos became human when Agrippa of Nettesheim and Leonardo da Vinci perceived the human form as pentagram.

The pentagram is the promise that a higher life is possible. But in the same way that being human inevitably involves error, the capacity to betray ourselves or turn things on their head, so the pentagram likewise loses all its glory when turned upside down. Like the human being, this geometrical form has to find its right relationship to its surroundings to be able to represent heaven on earth.

6

The Number of Perfection

And God saw everything that he had made, and
behold, it was very good. And there was evening
and there was morning, the sixth day.

Genesis 1:31

Perfection is the ordinary state of heaven

Johann Wolfgang von Goethe, *Maxims and Reflections*

Let's start with a little experiment, for anyone who has a microscope handy. It works best with a fine fall of snow, when ice crystals fall singly to the ground without clumping together into heavy flakes. If you catch single flakes on a cold sheet of glass and, keeping conditions cold, observe them under a simple microscope, something of the distinctive nature of six will be revealed. Every crystal, sometimes as compact as honeycomb, sometimes finely branching, will show a six-pointed star in millions of variations. More perfectly than in the growth of mineral crystals, these fleeting water crystals reveal and embody hexagonal symmetry; and one can observe this wealth of snow crystal forms for hours without ever tiring of them.

> The idea that, when snow falls in winter, something falls from the cosmos to the earth, is more than a merely poetic image. The number six, as expressed in this most archetypal phenomenon, is cosmic in nature.

Working with compasses, the radius of a circle can be transferred onto the circumference exactly six times. The circle, the simplest and most perfect geometric figure, seen in ancient times as the archetype of the cosmos, divides by six. And it is indeed the case that, from moons through planets to the stars and galaxies, almost

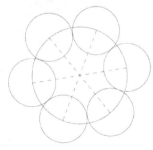

everything in the universe is round, rotates and thus bears six within it. It is scarcely surprising therefore that Babylonia, the great culture and epoch during which our spiritual relationship with the cosmos began to rupture, invoked six as the measure of all things. Wherever cyclical measurements come into play, in angular and time measurements (6 × 60 = 360 degrees, and 2 × 6 = 12 hours) the cosmic, Babylonian way of enumerating has been preserved. This mathematically cumbersome mode of counting and measuring has retained something of an inner connection with the cosmos.

The word cosmos comes from the Greek and means 'order' or 'ornament'. No other number possesses such a high degree of order as the number of the circle, six. It arises from its factors, and they constitute, as it were, its inner substance. The factors of six are 1, 2 and 3; and it is true only of six that the sum of these factors (1 + 2 + 3) as well as their product (1 × 2 × 3) both give rise to six again.

Just as the circumference of a circle leads us back to our starting point, so the factors of six lead us back to this same number. It rests in itself, and reproduces itself out of its parts. Or one might also say: the inside and outside of six are identical, and it is in absolute accord and congruence with itself. This is why Pythagoras regarded it as the most perfect number.

> Even if the rose produces the loveliest blossom, the most perfect bloom is the one without thorns, the lily, with its twice three petals. Imagined against a blue background, we can see its relationship to the whole universe.

Six is the focus of many religions. In the Persian religion of Zarathustra, there are six periods of Creation with six angelic beings. The six-pointed star, as an image of the interlacing of above and below, is present as the Vishnu triangle and Shiva triangle in Hinduism. And in Judaism and Christianity, too, six has a special meaning: for both in the Old and New Testament it stands for perfection: the world was created in six days; and on the sixth day at the sixth hour, the Son on the cross united himself with this world.

7

The Number of Time

Time is man's angel

Friedrich Schiller, *The Death of Wallenstein*

The Sumerians were the first to have a 'high culture of mathematics' from 3000 BC. Our 2 × 6 = 12 hours, and 6 × 60 = 360 degrees remind us that this ancient culture measured both time and space in terms of six. Why? Because the radius of a circle can be transposed to its circumference six times: as we saw, the number of perfection for a world created by and interwoven with the gods.

But then, four thousand years ago, the first world religion appeared which conceived God as elsewhere and transcendent. The earth as God's body became the world abandoned by God. With the advent of the Mosaic religion, the cyclical, ever-recurring measure in a divine world became a linear arrow of time as path towards a distant God. It was also this culture that embodied seven in the seven-day week, the festival of the seventh month, and the seven-armed candelabra, placing this number at the centre of its religious ideas.

What kind of number is seven? Like three and five, for instance, it has no factor and is thus a prime or original number. Multiples of 3 (6, 9) and 5 (10) are found within the first ten numbers. This is not true of seven. Already 14 lies beyond the numbers that can be counted on one's fingers. Seven is a number without affinities and can rightly be called the loneliest number because it neither contains other numbers within it nor can produce others.

Where do we find seven in nature? Apart from the seven-petalled Arctic Starflower, only flowers with four, five, six (or many) petals exist. Crystals are six or eight-faceted, but seven does not appear. Why not? The simplest explanation comes from mathematics. Heptagons cannot be used either to pave an area or to construct a regular geometrical solid – seven does not adapt to spatial conditions. And yet there is no realm of life without seven in a position of importance – usually where a process is involved, an enhancement or transformation. For example, there are seven colours in the rainbow uniting dark and light, and the cosmos of music arises from seven tones. The world's seven seas conduct water around the earth, and through seven orifices mammals and human beings see, hear, smell and taste the world, while seven organs sustain their life.

> Seven is the number of time – and accordingly, for example in the well-known tale of Snow White, the seven mountains across which Snow White passes to reach the seven dwarves indicate a complete transformation, comparable to the seven-year developmental phases in human biography, as the ancient Greek physician Hippocrates described them.

If seven is the number of transformation towards a divine state, it is hardly surprising that it also has high regard in Christianity: not only in the Book of Revelation, which refers to seven bowls of wrath, seven churches and seven trumpets (16:1, 1:4, 8:2), but also above all where transformation is apparent in the highest sense in death and resurrection. Holy Week, for instance, passes through seven phases that culminate in the seven sayings from the cross:

1. Father forgive them, for they know not what they do (Luke 23:34)
2. Today you will be with me in Paradise (Luke 23:43)
3. Behold your son ... behold your mother (John 19:26f)
4. My God, my God, why hast thou forsaken me? (Matt.27:46)
5. I thirst (John 19:28)
6. It is finished (John 19:30)
7. Father, into thy hands I commit my spirit! (Luke 23:46)

8

The New Number

There indeed arises the being of eight to which so far no one gave thought.

Johann Wolfgang von Goethe, *Faust, Part II*

1	א	*aleph* A		1	A α	*alpha*
2	ב	*beth* B, V		2	B β	*beta*
3	ג	*gimel* G		3	Γ γ	*gamma*
4	ד	*daleth* D		4	Δ δ	*delta*
5	ה	*he* H		5	E ε	*epsilon*
6	ו	*waw* W		6	F ϝ	*digamma*
7	ז	*zayin* Z		7	Z ζ	*zeta*
8	ח	*chet* Ch		8	H η	*eta*
9	ט	*teth* T		9	Θ θ	*theta*
10	י	*iod* I, Y		10	I ι	*iota*
20	כ	*kaph* K, Kh		20	K κ	*kappa*
30	ל	*lamed* L		30	Λ λ	*lambda*
40	מ	*mem* M		40	M μ	*mu*
50	נ	*nun* N		50	N ν	*nu*

Examples of Gematria for the Hebrew and Greek alphabets

If seven is the number of evolution, of the days in the week and the seven-year phases that bring us closer to our real being, eight leads us beyond ourselves.

The mathematician and philosopher Gottfried Leibniz wrote that 'mathematics is frozen music'. This may be why it is worth considering music when trying to understand numbers. The octave is the eighth note of the scale. Whereas the seventh makes a melody into a question, the octave is the answer or resolution in which the stream of tones comes to rest. The ancient view of the planets, similarly, saw the 'eighth sphere' as that of the fixed stars beyond the spatially and temporally accessible world of the seven classical planets. Eight, in other words, encompasses the sevenfold planetary system.

This encompassing quality of eight also comes to expression in many religions. The divinity Vishnu in Hinduism has eight arms in order to carry the world. Perhaps the best-known use of eight is in the Buddha's eightfold path. The term 'path' gives a false impression however. This core of Buddhist teaching should not be seen as a developmental path but as eight realms of inner transformation that stand alongside each other as equal aspects – from enhancing perception through right speech to meditation. Here too eight stands for the promise of another, higher world. The eight-rayed dharma wheel in the emblems of Tibet and Sri Lanka portrays the 'path' as symbolic image.

In Islam, there are eight levels of paradise, the Hasht-Bihisht, which signify redemption. Parks and gardens in Iran and India are therefore often laid out in eight parts.

The fact that a new world dawns with this number is also true at the elementary, mathematical level. Eight is the first cube number, since $2 \times 2 \times 2 = 8$. Space is embodied with eight: the cube, as most elementary solid of the spatial realm 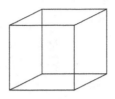 portrays this 'victory' in its eight corners. Mathematicians have also been preoccupied with why eight, specifically, is capable of forming squares of uneven numbers: $5 \times 5 = 3 \times 8 + 1$; $7 \times 7 = 6 \times 8 + 1$; etc.

A question more difficult to answer is why so many people are alarmed at eight-legged creatures, the spiders. It might well be because, with eight limbs, their blurring movement becomes impossible to calculate or predict.

In the Old and New Testaments, too, eight arises as a new, encompassing quality: mother and father, three sons and their wives enter the ark, and represent the foundation of a new humanity; and it is eight days after the entry into Jerusalem that the Resurrection occurs.

For number mystics something else is important too: in ancient Greece there was no distinct way of recording numbers, which were written down as letters. In alphabetical sequence, *alpha* to *theta* referred to numbers 1–9, *iota* to the now obsolete *koppa* 10–90, and *rho* to the obsolete *sampi* 100–900. Every name could thus also be transposed back into numbers. In Greek, the original language of the Gospels, the name Jesus was written as *Jesous*. Transformed into numbers the letters *iota, eta, sigma, omicron, ypsilon, sigma* give $10 + 8 + 200 + 70 + 400 + 200 = 888$. Jesus' saying, 'My kingdom is not of this world' could therefore be seen as an expression of the meaning and quality of the number eight. A triple number represents a strengthening of the single number, in this case the number eight.

9

The Almost Perfect Number

I know that I hung on a windy tree
Nine long nights,
Wounded with a spear, dedicated to Odin,
Myself to myself,
On that tree of which no man knows
From where its roots run.

The Edda / Hávamál

Ludwig van Beethoven, Anton Bruckner, Anton Dvorák and Gustav Mahler all completed nine symphonies. In nine books the Greek poet Hesiod tells of the nine muses, of whom Thalia, goddess of comedy and Urania, goddess of astronomy still impinge on our awareness today. What is nine, and what is its significance? The bigger a number the harder it seems to grasp its quality.

The nine is the last single-digit number, and is followed by ten. It thus marks the last step towards perfection. We not only find it amongst the composers we have mentioned but also in the world of astronomy: in the ancient ideas of Aristotle about the seven planets, the heaven of fixed stars and the unmoving movers enthroned above it, which give nine spheres surrounding the earth. In early Christianity, based on an apocryphal version of Genesis, there existed the idea of the nine hierarchies of angel choirs, starting with angels and archangels and extending to the Cherubim and Seraphim. It was primarily Dionysius the Areopagite, a student of Paul, who anchored this image of nine realms of angels in Christian belief.

This cosmic-spiritual context also exists in Islam. The ancient Persian astronomers divided the solar system into nine spheres.

Nine indicates a fullness that approaches perfection but, unlike seven, appears ungraspable and almost super-human. Tasks that exceed human capacity therefore frequently contain nine. Odysseus, for example, journeyed for nine years, Troy was besieged for nine years, and the annual festival of the Great Mysteries at Eleusis lasted nine days. The flood which Zeus sent to inundate the first human beings, from which Prometheus arose as the father of today's mankind, lasted the same length of time. We can find something similar in the northern myths: Odin, the highest Teutonic God, chained himself for nine nights to the mythical ash tree Yggdrasil, learned nine songs and in this way acquired the ability to write rune script. Here the nine appears as crisis moment and catharsis, out of which something new arises.

Nine – which is a lucky number in China – is contradictory. It is easy to calculate with it since the sum of the digits of any multiple of 9 produces 9 or a multiple of 9 (for instance, $49 \times 9 = 387$; the digits $3 + 8 + 7 = 18$ which is 2×9). At the same time nine is hard to grasp in geometric terms, since the nonagon, like the heptagon, cannot be constructed simply with a pair of compasses and a ruler. Nine is three times three, and three is the first number where we lose an overview. The three-body problem in physics or the wealth of interconnections between three people, are examples of the chaotic interplay that begins with the step from two to three. This wealth of ramifications multiplies further with the nine. Thus the etymological connection in many languages between 'nine' and 'new' (German *neun* and *neu;* or French *neuf* and *neuf)* is not accidental but expresses the fact that this number, like all that is new, cannot be immediately grasped.

> The most succinct account of the nine comes from Dante Alighieri, the Italian author of the *Divine Comedy*. In his *Vita Nuova*, he wrote of Beatrice, the idealistic and mystic love of his youth, who deeply informed all his creative work: 'This number [*nuova* or nine] was her true self.'

10

Grasping the World

Parents, teach how to love,
then you'll have no need of the Ten
Commandments – teach loving, I say.
In other words: be loving yourself!

Jean Paul, *Collected Works,* Vol. 121

Our hands are busy whenever we write, knit, draw, serve food, tie knots or repair things. And our eyes follow the movements made by the hands. The gaze focuses on our two hands more than any other part of the body. Looking at our own face or whole form in the mirror strikes us as far less natural than observing what our hands are doing; for whenever our hands are at work – drawing, cooking, typing or modelling – we are also engaging with the world, or doing something for it. In this 'grasp of the world' the number ten plays a natural role corresponding to our ten fingers. Throughout life we visibly grasp hold of the world with ten. And this experience has helped the decimal system to its victory over other counting systems originating in ancient times. There may, for instance, be good grounds for using eleven different numbers and only reaching a new numerical starting point with twelve, a duo-decimal system; but our ten fingers gave the decimal system its edge over the others.

In the school of Pythagoras, ten was called the 'all-encompassing mother', for it is the sum of the first four numbers, the *tetraktys*: to

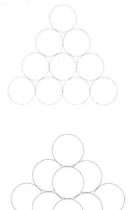

arrange circles in a triangle, place two below the top one, then three below that, followed by four. The sum of the circles thus grows from 1 to 3 to 6 and finally to 10. These four steps lead to ten, and having arrived there, everything is expressed: one indicates the world, two the polarity of phenomena, three the spirit, and four the earth.

Ten is not just a 'triangular number' but also a 'tetrahedral number'. In balancing balls on top of each other to form a pyramid, you first make a base of six, then place three on top, and finally one above them, so that ten arises once more as totality.

In numerous cultures and religions the ten fingers reappear as ten rules or commandments through which to engage with the world. The most significant are the Ten Commandments in the Old Testament which, in a novel about their origin, Thomas Mann calls the 'ABC of human conduct'. Christianity adopted the Ten Commandments, though different Christian churches relate to them in different ways.

In the Qur'an (sura 17), picking up on the Old Testament tradition, there are also ten commandments; and even in Buddhism there are ten rules, or rather obligations taken voluntarily: five for the faithful, such as the duty to protect the lives of others, and five additional ones for monks and nuns, such as the duty not to accept gold or silver.

Interestingly, in the twentieth century, ten-part ethical frameworks have been established where religious belief is rejected, yet where an ethical and moral stance is sought. Thus in the first years of the Second World War, German soldiers were issued with '10 rules of warfare' (ethical guidelines similar to the Geneva Convention). In the former (east) German Democratic Republic, the Socialist Unity Party formulated 'Ten rules of socialist morality'; and 'Ten rules for the Young Pioneers' were printed on membership cards of the GDR's youth organisation.

The figure 10 for ten makes sense because, as all-encompassing number, it is related to the number one – it might even be called the 'daughter of one'. Thus Paul said that if one practised love of one's fellow human beings one would not need the Ten Commandments. Seventeen hundred years later, the writer and educationalist Jean Paul urged parents to take note of the same thing.

11

Crisis and Bridge

All transitions are crises,
And isn't a crisis an illness?

Johann Wolfgang von Goethe, *Wilhelm Meister's Apprenticeship*

Eleven's neighbours are great numbers, for both ten and twelve appear wherever wholeness is indicated. But what does this mean for eleven as the gap between these two glorious majesties?

'Eleven is an evil number. The zodiac has twelve signs. Eleven is sin and goes beyond the Ten Commandments' says the astrologer in Friedrich Schiller's play *Wallenstein*. This corresponds to the medieval idea that a quantity possessing eleven parts, or a period of time lasting eleven years, is *ad malam partem*, or in other words must be regarded negatively.

Franz Kafka's story 'Eleven Sons' picks up on this. It tells of a father who complains about all his eleven sons because he can only see the negative in each. Eleven exceeds ten, the number of order and the law, yet without attaining to twelve, the number of perfection. A number of crisis therefore? This 'beyond ten' is also what gave eleven its name, derived from Old English *endleofan*, literally 'one left over'.

Is this 'one too much' really an appendage that drags eleven down? There is much to suggest this is not the case. The best-known story about eleven in Germany comes from Cologne and can still be seen in that city's emblem as 'eleven flames'. This is the legend of the Breton princess Ursula who undertook a pilgrimage to Rome with ten companions. On their return journey on the Rhine, in the year 452, these 'eleven virgins' fell victim to massacre by the Huns in Cologne. In memory of this legend, Columbus named a group of islands in the Caribbean the 'Virgin Islands'. In this story, eleven is not of less worth than the round number ten, but is an enhancement of that number by the princess.

In ancient Athens, the court of judgment was composed of eleven men. The goalkeeper in modern football matches likewise has a special position and therefore (usually) bears the number 1 on his shirt. He enables the ten other players to look ahead. (A penalty kick, though, is taken not from eleven metres, but from twelve yards outside the goal.)

But where do we find the intrinsic quality of eleven? In the following mathematical experiment for example: make a cube of paper by cutting out six conjoined squares then folding and gluing them. Interestingly, there are eleven ways of joining six square surfaces together to produce a cube. By eleven different routes, therefore, one can pass from two to three dimensions. Here eleven is not just a gap, but also a bridge.

The best-known natural expression of this number is probably the eleven-year sunspot cycle. Every eleven years on average, the sun is populated by numerous local darkenings that affect electromagnetic storms that cause the aurora to appear, and inhibit plant growth. We still do not fully understand this solar phenomenon, nor why at present there are fewer discernible sunspots than for a century. They are, at any rate, striking disturbances of the sun's light and heat. 'The bright sun brings it to light' (the title of a Grimm's fairytale) can, if we relate it to eleven, suggest that this number is one of transition and crisis.

This also accentuates the aspect of the bridge, which is at the same time an image of crisis: we leave an old shore behind to find our way to a new one. Mathematically, these shores are ten and twelve.

12

The Number of the Whole World

And he ordained twelve, that they should be with
him, and that he might send them forth to preach.

Mark 3:14

Eleven members of the jury are in agreement, for witness statements clearly show the crime was done by the 18-year-old Puerto Rican. Only the twelfth jury member is still pondering, and refrains from voting in the first round, while the other eleven all declare the youth guilty. It is precisely the apparent clarity of the case that gives one person pause for thought. He wishes to think it through slowly, rather than jumping to a conclusion; and gradually reveals how all the other eleven jury members were biased and wrong – each in their own way. In 1957 the film *12 Angry Men* was first shown in cinemas. It is one of the most revealing court-scene dramas because it shows, in striking fashion, that twelve voices are required for a fully responsible social judgment – that the circle is only complete with twelve. Indeed, if the spirit (three) and earth (four) are multiplied together – which means to engage in the greatest possible exchange with one another – twelve results as a number that represents the whole.

> This idea that twelve is an expression of totality, and a path towards it, is present in many cultures: for instance the twelve Greek gods of Olympus, or the twelve labours of Hercules.

The further back we go, the greater does twelve appear. In the Old Testament, for example, the people of Israel are composed of twelve tribes. Here too they are eleven initially, but eventually Rachel's pleas are heard, and she gives birth to Benjamin (the youngest) as twelfth tribal father. Twelve disciples gather at the Last Supper and represent all humanity, reflecting on earth the circle of twelve zodiac images in the heavens. Actually, these signs were also originally eleven in Babylonian times, for Libra, the Scales, was seen as the claws of Scorpio. In the fourth century BC it acquired autonomy as a separate, twelfth sign, in accordance with the twelve months of the lunar year.

> We find the first epic, the story of King Gilgamesh of Uruk – the capital of Mesopotamia – recorded on twelve clay tablets that were discovered in the nineteenth century. This is the great story of mortality, love and friendship between Gilgamesh and his companion Enkidu, the wild man of the forest – who ultimately dies after twelve days of illness. Here again there were originally eleven clay tablets in cuneiform script. A subsequent twelfth tells how Enkidu once more comes back to Gilgamesh from the underworld, thus conquering death.

Whether as the twelve ribs which encompass the chest, or the twelve semi-tones from which music arises, a dozen always indicates the whole. This phenomenon extends right into physics. Jupiter, which passes through the zodiac in twelve years and whose diameter is twelve times that of the earth, is the planet with the biggest circumference. Any additional matter that it absorbs increases its density without adding to its size.

The finest expression of twelve is in human biography. At the age of about twelve children start discovering natural and mathematical laws, and thinking autonomously. Justice now becomes an important concept. Abstract thinking awakens and with it the capacity to encompass the whole world in thought.

13

The Step into the Unknown

The hope that evades risk is no hope.
Hope means believing in the adventure of love,
having trust in people,
taking a step into the unknown and giving oneself
over to God.

Dom Hélder Pessoa Câmara

It is the excess number, the step into the unknown, the first number after the completion of twelve. It is the one-too-many that exceeds the dozen, and also the first number whose name in many languages is composed of two numbers, three and ten. As such it is harder to grasp than its numerical predecessor and therefore deserves the epithet given to it in a children's book by Michael Ende, *Jim Button and the Wild 13*. There are two main reasons why 13 has been thought an unlucky number since the seventeenth century, so that hotels often do not have a thirteenth floor. Firstly because at the Last Supper one of the thirteen betrayed Jesus, and secondly because on Friday October 13, 1307, Philippe IV, king of France had all Templar knights arrested and annulled the Order. The Templars were the first community to unite knighthood with a monastic order, and they gained enormous influence as protective force during the Crusades.

> In Italy though, the Friday that falls on the seventeenth of the month is regarded as the unlucky day. This is because the Roman figures for 17 (XVII), if rearranged, give the Latin word *vixi*, meaning 'I have lived' (or 'life is over').

Even if we take a step into the unknown with 13, in mathematical terms an ordering power issues from this number. Thus one can find 13 axes of symmetry in a cube – the simplest geometric solid. Three axes pass through the planes lying opposite to each other, four axes as diagonals connect each pair of the cube's corners, and six axes pass through

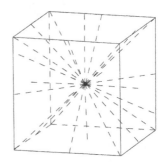

the centre of opposite edges. With a cube of cheese and 13 toothpicks one can build this configuration; or undertake it as geometric meditation: imagine you are at the centre of the cube, with your feet and head touching two planes, your outstretched arms touching the planes to left and right, and behind and in front of you the other two planes. Now you can incorporate one axis after another into this imaginary cube, until you have all 13. A difficult concentration and spatial picturing task.

There is a further realm where the 13 orders space, for there are 13 semi-regular or 'Archimedean solids'. Unlike the Platonic solids, these are not composed of a particular regular surface but of different ones. The best known of these solids is the classic football, the so-called 'truncated icosahedron' formed of pentagons and hexagons. To be able to picture all 13 of these solids in your inner eye would certainly be a hard geometrical exercise.

A division into 13 is also found in the path of sun and moon: each season consists of 13 weeks, and there are 13 days from the appearance of a new moon to full moon.

Thirteen as ordering number can also be found in the Old Testament. In Exodus, 13 attributes of God are invoked, starting with 'The LORD, the LORD, a God merciful and gracious, slow to anger, and abounding in steadfast love and faithfulness' (34:6f). There may even be a connection here with the fact that in number mysticism the Hebrew word 'one' *(ahad)* has the numerical value of 13, so the oneness of God encompasses a 13. In the Kabbala, the book of Jewish mysticism, 13 heavenly springs, 13 gates of grace and 13 streams of balm are said to await us in the life after death. In the Zohar, the key book of this teaching, it is said that God may be found by pursuing '13 paths of love'.

14

The Bridge Between Heaven and Earth

Only when ripe
Does the fruit of destiny fall.

Friedrich Schiller, *The Maid of Orleans*

It takes 14 days for chicks to hatch from the eggs of blackbird, blue-tit and sparrow, swallow, starling and many other birds. For well-known reasons, the cuckoo takes two days less. The length of nesting for most birds is geared to the rhythm of the week, starting with around 14 days. Other birds grow in the egg for about 21 or 28 days.

The number 14 is twice seven, the second step in the seven-times table, and therefore, like that number, connected with time and development.

This is particularly apparent in human biography. School in Germany, Scandinavia and in Waldorf Schools begins during a child's seventh year – that is, between the age of six and seven, when physical development is far enough advanced to allow intellectual development to begin, with writing, reading and arithmetic. At age 14 – or somewhat earlier today given good nutrition – soul maturity succeeds physical maturation, awakening in love for the opposite sex but also in enthusiasm for ideas and ideals. From being an 'offspring', one becomes a member of humankind. Thus it takes 14 years, or 14 steps, to come to belong fully to the human race. It is perhaps due to this that in the sixteenth century the idea formed of the 14 Stations of the Cross on the path to the Crucifixion. The sentencing of Christ, shouldering the cross and the first fall are the first three stations, while the fourteenth is the entombment of the body. Though enhanced in a great mythical image, this path also corresponds to the goal of becoming human. Another Christian tradition seems to be derived from this: the Catholic idea of the 14 holy helpers – saints and martyrs of the early Christian period to whom one can appeal at times of dire need.

Fourteen also surfaces as an important factor in the world of physics, in a realm quite opposite to the spiritual path of development. The French mathematician Auguste Bravais

Table salt

investigated the different means by which the tiniest particles in space can arrange themselves to form a crystal lattice. Table salt, for instance, forms a so-called 'cubic face-centred lattice'. This means that the smallest structure is cube-shaped, with one atom respectively situated not only at the eight corners of such a conceived cube but also at the centre of the six faces or planes. Accordingly, a cube of this kind has 14 particles.

There are other possibilities however: instead of the face centres, the edge centres can each have one atom, or the distances can be extended in one direction so that the structure becomes oblong. Through such variations in crystal formation, a total of 14 different spatial structures are possible – 14 different so-called 'Bravais lattices' (see p. 106).

Fourteen is not a temporal factor here but, interestingly, determines the multiplicity of crystalline structure.

In Islamic number mysticism, this breadth of the number 14 is well known. There it is regarded as the number of the moon, since there are just about 14 days between new moon and full moon. It determines the rhythm in which the moon, as the bridge between earth and cosmos, grows to full size and fades away again to nothing. Thus 14 is a mediating rhythm between heaven and earth. Crystals, with their special relationship to the light, likewise stand in this heavenly-earthly axis, as do the birds we mentioned at the start.

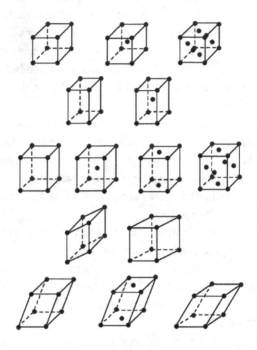

14 Bravais lattices

Melancholia *by Dürer*

15

The Underestimated Number

An ancient Chinese legend tells of a turtle emerging from a flood with a curious pattern on its shell: circular dots of numbers arranged in a three-by-three grid pattern such that the sum of numbers in each row, column and diagonal was always the same: 15, which is also the number of days in each of the 24 cycles of the Chinese solar year.

4 9 2
3 5 7
8 1 6

In the mysterious picture by Albrecht Dürer entitled *Melancholia*, it hangs on the wall behind the pondering angel. In Dan Brown's novel, *The Lost Symbol*, it is a secret message; and there is a theory that Goethe's 'Witch's one-times-one' in *Faust* is an instruction for creating the 'magic square': numbers arranged in a square in such a way that, added together, the sums in all directions give the same total. Thus the different numbers are incorporated into a harmonious whole that works out precisely like an all-embracing mathematical formula.

The simplest form of the magic square is the Chinese *Lo Shu*: the nine basic numbers are arranged in a three-column square in such a way that 15 always results. A Chinese story tells how the mythical Emperor Fu Xi was meditating by the river, from which emerged a tortoise whose shell bore this magic pattern. In Taoism the square plays an interesting role, symbolising the cosmic balance between the opposites 'yin' and 'yang'. The even numbers (image of the feminine) stand in each of the four corners, while the odd numbers (symbol of the masculine) are in the middle of each side. In the middle is five as number of the five Chinese elements and compass directions (north, south, east, west, centre). Thus 15 appears as the number of equilibrium.

Every number can also appear in apparently random places. In the case of 15, for instance, we can find this in tennis scoring: steps of 15 rather than 1, 2, 3 (as in table tennis) giving 15, 30 and (originally) 45. There are historical reasons for this awkward scoring method: around 400 years ago the forerunner of tennis, the French *jeu de paume* in which the ball was struck with the flat of the hand, was usually played for money. The winnings per point were 15 deniers. The 15 balls in pool billiards also probably have no deeper reason than this.

Numbers seem to resemble people: whereas great individuals are able to reveal distinctive aspects of their character and being, the real qualities of many of us are displaced or concealed, and can only be grasped at one remove. Fifteen, similarly, only seems to reveal something of its intrinsic nature in unexpected places. In the Bible, whenever a number is mentioned, this often indicates a distinctive characteristic, a quality referred to as quantity. The amount is not chance, especially not in such a central passage as the hymn to love by St Paul in his Letter to the Corinthians which starts with the much-quoted line: 'If I speak with the tongues of men and of angels, but have not love ...' (1Cor.13:1). This is followed by the invocation of fifteen, enormously far-reaching attributes of human love:

> Love is patient and kind; love is not jealous or boastful; it is not arrogant or rude.
> Love does not insist on its own way; it is not irritable or resentful; it does not rejoice at wrong, but rejoices in the right.
> Love bears all things, believes all things, hopes all things, endures all things. (1Cor.13:4–7)

16

Ordering the World

Life is breathing order.
Order is remembered love.

Peter Horton (Austrian singer and musician),
Die zweite Saite ('The Second String')

A hand-breadth is the width of four fingers, and four hand-breadths correspond to the length of the foot. Thus 16 fingers make one foot. In ancient Rome as in Greece, a system of measures was based on this. This also seems to be why in all Romance languages each number up to 16 has its own name, with compound names starting only from 17: in French *seize* (16) is followed by the two-number *dix-sept* (17); and in Italian *sedici* (16) is likewise followed by the compound *diciasette* (17).

What is the mathematical quality of 16? It is the first square of a square: $2 \times 2 = 4$ and $4 \times 4 = 16$. By squaring numbers a distance becomes an area. Whereas a distance only possesses length without breadth, and can therefore only be understood in one dimension, an area extends both lengthwise and breadthwise.

Four is therefore more earthly than two, and 16 more earthly than four. This may be the underlying reason for 16 as foot-based measuring system, since this number has a particular relationship to earthly conditions. Of course there are other numbers that are squares of squares: 81 $(3 \times 3) \times (3 \times 3)$ is the next, followed by 256 $(4 \times 4) \times (4 \times 4)$. But only in the case of 16 can number and power be interchanged $(4^2 = 2^4 = 16)$. In the case of 81 this isn't possible, for 3^4 gives 81, whereas changing the numbers round to 4^3 gives 64.

Sixteen – like twelve – shows a special kind of perfection, which is why it was used in ancient cultures as an index measure. In spatial terms the 16 division can be found in the compass rose, with directions passing, for example, from south to south-southeast to southeast to east-southeast, etc. As time measure, too, 16 forms a whole composed of 16 parts: a simple melody – such as most children's songs (like Ba-Ba Black Sheep) – lasts for four bars, each of which has four beats. In the classical Indian rhythm of *tintal,* one finds this in still purer form, with 16 beats in a bar.

In India, until 1957, the rupee was divided into 16 annas; and in the Indian Vedas it is said that the human being is formed of 16 parts. The ancient Indian divinity *Pussa* was pictured as having 16 arms.

Sixteen is also to be found in European culture. The Greek doctrine of the four elements of earth, water, air and fire was expanded in Rosicrucian alchemical teachings into the 16 so-called philosophic elements, by connecting each of the four elements with the others; as in: the fire of earth, the fire of water, the fire of air, the fire of fire etc.

Such a spiritual or philosophical way of perceiving the spirit in nature is no doubt hard to understand today. But most people probably have a personal relationship to 16 when they look back to their youth, for at this age, often, love of another reaches a first culmination – and here again 16 appears as image of wholeness. Love of someone starts with the discovery of one's own one-sidedness, and grows through longing towards fulfilment.

17

The Loveliest Number

Beauty is the memory
Of the spirit's power to conceive.

Joachim Daniel, lecture 'On the Muses' at the
Anderzeit Drey conference, 2009

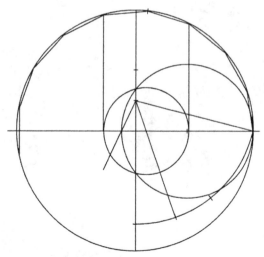

Heptadecagon construction

For over two thousand years – well into the nineteenth century – the best-known book after the Bible was *The Elements* by the Greek philosopher and mathematician Euclid. This 13-volume work contains the mathematical and geometric knowledge of ancient times, and almost two millennia passed before something new was added to it. This new knowledge, this latter-day answer to Greek geometry, was the number 17. The 19-year-old mathematician Friedrich Gauss succeeded in demonstrating how and why the heptadecagon (17 sided figure) could be constructed solely with a pair of compasses and a ruler. Whereas heptagon (7), nonagon (9) and undecagon (11) cannot be formed by this simplest of means, it is possible to construct a heptadecagon without protractor and set square. One has to draw 47 reference lines, then add a further 17 lines before finally – like a mountaineer triumphantly reaching a summit – allowing this special polygon to appear on the paper.

Gauss had not discovered a mathematical peculiarity, something merely odd and exotic, but had succeeded in delving deeper into the structure of numbers through 17. Like 3, 5 or 257, 17 is a Fermat number, named after the French mathematician Pierre de Fermat. These are numbers which directly follow the especially simple numbers – 2×2 (= 4), $2 \times 2 \times 2 \times 2$ (= 16) – and therefore, even though they are prime numbers, can be constructed as polygons. Although they are hard to grasp as prime numbers, they take something from their simple, straightforward predecessors and can therefore be conceived as outline or shape. They are paradoxical numbers because (as prime numbers) they are both tangible and yet also intangible.

The memorial to Friedrich Gauss in Braunschweig therefore has a golden heptadecagon at the feet of this 'prince of mathematics'.

The mathematical peculiarity of 17 cannot however fully resolve why, when asked for their favourite number between 1 and 20, over 70 percent of people name this number. What makes 17 the 'loveliest' number?

Modern language and music both take their point of departure from 17. In the Greek sequence of speech sounds, the first complete alphabet, the seven vowels are accompanied by 17 consonants or phonemes that lend language its structure. The same is true of music. In classical Greece likewise, Pythagoras discovered the mathematical structure of the musical intervals. To advance one tone higher on a stringed instrument, we must shorten the vibrating string length by a ninth. If the original tone has nine spans, the next higher tone has only eight. Together this gives $9 + 8 = 17$ spans. Thus 17 contains the measure of progression of the well-tempered musical scale.

Now it is interesting that 17 appears in myth precisely where the greatest steps – those into the unknown – are taken. In ancient Egypt, Osiris in his coffin was given over to the floodwaters of the river Typhon on the seventeenth day. The Flood also began for Noah on the seventeenth day of the month, and ended again, likewise, on the seventeenth day of the seventh month. It is the number of transition – not least in human life where it signifies the passage into autonomy and freedom.

18

Life in its Fullness

Responsibility for oneself
Is the root of all responsibility

Mencius

Like 12 or 24, 18 is a 'rich' number in the Greek sense. This attribute is given to numbers when the sum of their factors exceeds the original number. Their inner structure enables them, as it were, to grow beyond themselves. The factors of 18 are 1, 2, 3, 6 and 9, which together add up to 21. This means that the number 18 contains more than it shows outwardly.

> The inner power of 18 can also be found in Jewish numerology. Since every number is written as letters, each can be regarded as a word. Eighteen is written as *chet, yod,* together forming the word *chay,* which means 'life'. Students of ancient customs suspect that this connection in Judaism between 18 and life is also why the core Jewish prayer, which is spoken thrice daily, focuses on 18. The prayer is called Shmone Esre, which means 18, and contains 18 pleas or petitions.

This fullness of life embodied in 18 is present in the only passage in the Bible that mentions this number: 'And a woman was there who had been crippled by a spirit for eighteen years. She was bent over and could not straighten up at all.' Here too the number is not random. It is the image of 10 and 8, of law and grace, but certainly also an expression of life's fullness.

In Islam, too, 18 occupies a key position. The introductory phrase that prefaces the suras of the Qur'an runs: *Bismillah al Rahman al Rahim* (in the name of God, the Most Gracious, the Ever Merciful). This consists of 18 Arabic consonants. Speech comes to consciousness in the consonants, and these 18 consonants therefore signify 18 moments of awakening.

A special significance is accorded to 18 in Islamic mysticism. The opening poem by the great poet and mystic Rumi, the 'Song of the Reed', has 18 verses. In accordance with this poem, all who wished to become a whirling dervish had to serve for 18 days in the monastery and perform 18 kinds of kitchen service – ranging from bread baking through washing dishes to cutting vegetables. After proving their worldly aptitude in this way, spiritual service followed: the adept was led with an 18-armed candelabrum into a cell where he had to examine himself in 18 days of meditation.

In Buddhism, the expression '18 arms of the Buddha' has special significance originating with Bodhidarma, the first spiritual leader of Zen Buddhism. At the Shaolin Temple in central China he developed physical exercises to help people endure long hours of sitting still in meditation. The best-known sequence of exercises was called 'shiu-ba-lo-han-shou' (= the 18 hands of the Buddha). Later these exercises were found useful for self-defence, so that the hands of the Buddha developed into types of martial art such as Kung Fu.

In the macrocosm of the universe, we again find 18 as the rhythm of a certain type of life. Solar eclipses are the most striking manifestation of life's interruption. The special configurations in which the moon inscribes bands of eclipse upon the earth recur in a rhythm of 18.03 years. Since there are 42 interpenetrating 18-year eclipse cycles of this kind, two to three solar eclipses occurring each year.

Our strongest relationship to 18 lies of course in the advent of adulthood. On 20 November 1989, the UN General Assembly concluded the Convention of Children's Rights, according to which the age of majority is attained at the age of 18. Except for a few African and Arab states, where 19, 20 or 21 marks the arrival of adulthood, the age of 18 has universally come to signify the end of childhood, and the step into autonomous life.

19

The Number of New Birth

We come into the world
In order to be born

Pablo Neruda

Each year, on March 12 or 13, always at 8.52 am, and in the autumn on September 30 or October 1, always at 9.33 am, an alpine lightshow takes place at the little Swiss mountain village of Elm. As the day dawns, a brightly glittering light shoots out from the eastern wall of the mountain: sunlight finds a path through a hole, the Martin's Hole, in the cliffs, and casts a circle of light upon the village. The light progresses at 32 cm (13 in) per second. Every 19 years the Elm occurrence intensifies, for not only does the sun shine in this way in the morning, but in the evening the full moon also sends its light through the opening in the cliff. The last such event was in 2009, so we'll have to wait until 2028 to see the full moon in the Martin's Hole again.

This rhythm is due to the fact that it takes 19 years for the full moon to re-establish the same relationship with the sun on the same date. In ancient times, the Greek mathematician Meton discovered that 19 years corresponds precisely to 235 moon cycles during which the moon's phases return to the original date. Thus a full moon occurred at Christmas Eve 2015 and will not fall on the same date again until 2034. Since the Christian calendar is oriented to the course of the sun, whereas the Islamic calendar relates to the moon, these two types of time reckoning also resonate together in a rhythm of 19 years.

But as well as sun and moon, the earth and moon likewise resonate together in a recurring rhythm of almost 19 years. The lunar orbit is inclined to the earth's ecliptic at an angle of 5 degrees. This angled plane dances round the earth like a hula-hoop. After almost 19 years, or more precisely 18 ⅔ years, the slant of the lunar orbit faces the same direction again. As long ago as 3000 BC, this moon rhythm was traced by means of ancient stone circles.

This so-called 'moon-node' rhythm plays an important part in our biography. After half this period, 9 ⅓ years, when the moon orbit is opposite to what it was at the time of birth, we also reach the greatest distance from our birth impulse. Experiences of loneliness and exclusion are typical at this edge of autonomy. A friend at this threshold between childhood and youth once confided to me: 'I feel that no one else is real. The gods put them on earth to test me.' When the moon's orbit returns to its original position again after almost 19 years, a moment of spiritual birth has arrived once more. Whether through crises, illness or important encounters, a new point of departure is due, a further birth. From a spiritual perspective the impulses and tasks we took upon ourselves before birth are renewed. For this reason too, multiples of 19 (or rather 18 ⅔) years mark biographical turning points. Nineteen is the number of birth and of the moon that leads us towards birth. The ancient Greeks called earth's satellite, 'cradle' and 'grave'.

If we count the consonants of the alphabet according to their sound, or in other words equate *c* with *k,* and *v* with *f,* (these two sounds are identical in German) then this gives 19 phonemes. Whereas the vowels a or o, as 'ah' and 'oh', accentuate our emotional response to the world, the consonants lead to a sense of self – which again is a birth motif. It is likely that 19's close relationship with birth and beginning is the reason why, apart from seven, scarcely any other number figures so often in popular song lyrics: from Paul Hardcastle's *Nineteen* (because of the average age of the soldiers sent to Vietnam) through Steely Dan and his *Hey, Nineteen* to the Rolling Stones with their legendary *Nineteenth Nervous Breakdown* or Joe Jackson's *Nineteen Forever.*

20

Space and the Human Being

At twenty years of age the will reigns;
at thirty, the wit, and at forty, judgment.

Benjamin Franklin, *Autobiography*

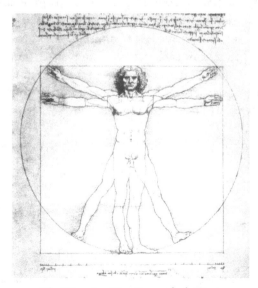

The Vitruvian Man, *Leonardo da Vinci*

Called *alanine* and *arginine, leucine, lysine* and *tyrosine*, they are the constituents of life. The twenty alpha amino acids are such that a new molecule can join on to the end of every amino acid molecule. When two amino acids combine they are called 'dipeptides', and when several join together, 'polypeptides'. Where over 100 combine, protein, the stuff of life, is formed. At 16 percent of body mass, it is the most commonly occurring substance in us after water. Muscles and tissue consist chiefly of protein, which dictates bodily form and, with a host of enzymes, also governs organic processes in animals and human beings.

We can see something of the mystery of proteins if we take a look at a fried egg in a pan: just a little heat is enough to render the protein – or rather the proteinous white of the egg – opaque. What's happening here? The protein becomes denatured and changes its spatial structure. Protein is distinguished by its complex convolution which, depending on the particular amino acids joined together, gives rise to a unique microcosmic cluster – which is different in each person. The phrase 'immunological individuality' refers to distinctive protein structures whose fingerprint consists of the special convolution of protein formed from an interplay of the twenty amino acids.

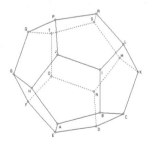

Dodecahedron

Icosahedron

Here, and also in geometry, twenty gives rise to spatial structure. Twenty is the highest number to appear in elementary form in the fully regular Platonic solids. Thus the dodecahedron formed of twelve pentagons possesses twenty corners, while the icosahedron related to it has twenty triangular faces. No larger number than twenty appears in the regular solids.

Twenty appears in its simplest, most tangible form however in hands and feet. Twenty toes and fingers point out into the world. It is scarcely surprising therefore that many cultures, such as the Mayan, have a counting system based on twenty. Starting with *xix im* for one, through to *kal* for twenty, they used a vigesimal system that counts up to twenty then uses multiples of twenty. We find something similar in the ancient Irish language and in French, where eighty is called *ceithre fichid* or *quatre-vingt* respectively; 4×20. The same is true of Danish, where the word *firs* (shortened from *firsindstyve*) also means 4×20; and seventy is *halvfjerds*, meaning $3 \frac{1}{2} \times 20$.

In our twenty toes and fingers, and also in our twenty milk teeth, in geometric solids and the twenty components of protein, this number always embodies a grasp and forming of space.

The factors of a number often tell us something about its character. Four and five are factors of twenty. The four directions of the compass and the four elements reveal four's earthly, spatial nature. Five appears in the ancient lore of the elements as the *Quinta Essentia,* the spiritual nature of the universe and the human being. In his painting of the human figure within the pentagram, Leonardo da Vinci depicted this relationship with five. Twenty is the product of these factors and thus represents the interplay of space and human being.

This may be why, in Japan the age of adulthood comes at twenty. Over this period of time a human being unites with the earth sufficiently to be able to take full responsibility for themselves.

21

The Number Between
Eternity and Time

Time is like eternity and eternity like time
As long as you yourself do not
Make different what is the same.

Angelus Silesius, *The Cherubinic Wanderer*

The cannons thundered 21 times when the Crown Princess of Sweden said 'I do' in June 2010, and 21 times when President Obama was sworn in. And there is a 21-gun salute when the Queen Elizabeth visits a foreign state. Why 21?

When sailing ships were equipped with cannon in the fourteenth century, they were not allowed to enter harbour primed for battle. They therefore fired off the powder from their cannon without cannon balls, and this sign of peace led to the gun salute. Seven such shots were fired since these early battle ships each had seven cannon. Possibly because cannon on land could fire more quickly, the seven salute shots became 3 × 7 shots, and have remained so to this day.

But there are also deeper reasons for this twenty-one-fold thunder: twenty-one is a grand number, less in terms of quantity than due to its inner properties. It is the product of three and seven. Beside the three totals produced from opposite numbers when playing dice (1 + 6, 2 + 5, 3 + 4 = 21), three and seven are also both powerful numbers in their own right. Three is the number of the spirit, of the divine Trinity, and seven the number of development, of time. Thus 21 is the union of spirit and time. At the age of 21 the human spirit stands fully within time, or in other words in the earthly here and now, and has reached the age of majority in the fullest sense.

Whenever an enumerated list appears in religious texts, the number invoked is rarely chance. Probably the biggest single list that appears in the Bible itemises 21 qualities of wisdom in the Book of Wisdom, in the Apocrypha. These are, one can say, a 21-gun salute as hymn of praise to the divine value of wisdom. A spirit is said to live in wisdom that is 'intelligent, holy, unique, manifold, subtle, mobile, clear, unpolluted, distinct, invulnerable, loving the good, keen, irresistible, beneficent, humane, steadfast, sure, free from anxiety, all-powerful, overseeing all, and penetrating through all spirits that are intelligent and pure and most subtle'. (Wisd.7:22f) The human capacity for wisdom is nothing other than the faculty for drawing from the world of ideas, the eternal world, something that fills language and memory and is thus introduced into temporal conditions. Eternity is brought into time through the interplay of seven and three, the two factors of 21.

> In Tarot cards widely used to interpret or predict destiny, the twenty-first card is highest in worth, and represents the world or the whole universe.

A significance that 21 had in ancient times is that of being the triangular number of six. What does this mean? Whereas we reach the number six by adding one six times, in the case of triangular numbers we add first one, then two, then three, giving rise to a special numerical series which includes the distinctive numbers 36, 153 and 666. The sixth step in this series gives 21, for $1 + 2 + 3 + 4 + 5 + 6 = 21$.

Thus 21 is a higher expression of six, which is a number of great perfection.

In our current century, any culture whose calendar is based on the Christian time reckoning still has the remaining years of the twenty-first century left to allow 21 to become fully part of our language and understanding. It is clear from the special relationship that this number has to time and eternity that this is a worthwhile endeavour.

22

The Number of the Sun

Nothing more beautiful under the sun
than being under the sun ...

Ingeborg Bachmann, *'To the Sun'*

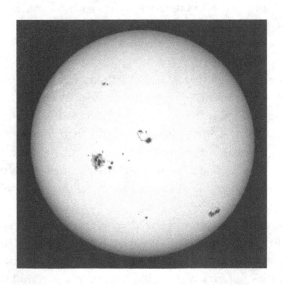

The next time sunspot activity is expected to reach its greatest extent is sometime around 2025. Sunspots can be larger in size than the earth. In the first century BC, Chinese astronomers were the first to detect these mysterious spots on the setting sun. It must have been a shock to ancient peoples to find that the central star, the symbol of love and wisdom in most religions, this 'natural phenomenon most closely resembling God', had impurities.

If we look at the sun through a filtered telescope – whose invention in the seventeenth century first allowed sunspot research to be conducted regularly – these spot-like formations 1,000–50,000 km (620–31,070 miles) in diameter look like black occlusions. The dark colour is deceptive however, for even inside these spots the temperature is still 4000°C (7232°F), emitting correspondingly bright rays. Compared with their surroundings though, whose temperature is 6000°C (10832°F), the spots appear dark and cold. Violent magnetic vortices within them exclude the influx of 'fresh' heat and light from the sun's interior.

After a few weeks or months, the spots fade again, while new ones form elsewhere.

The frequency of their appearance follows a long-wave rhythm. Every eleven years on average, the number of sunspots reaches a maximum. This sunspot cycle is an expression of a turbulent inner process within the sun. It is still little understood, but leads to a reversal of the sun's whole magnetic field every eleven years. So, after 22 years, the magnetic field returns to its previous orientation, and so the sun's intrinsic cycle, the Hale cycle, lasts 22 years.

To perceive this and observe the appearance of sunspots one does not necessarily need a telescope. If you watch the aurora in high latitudes or, as meteorologist, take an interest in the density of the earth's outermost atmospheric skin, you can find the same eleven-year rhythm underlying the 22-year rhythm as a $\frac{2}{4}$ pulse. Like the northern lights, the density of the exosphere is directly connected with the sunspots. The earth, one can say, really breathes in the rhythm of the sunspot cycle.

It is interesting to see where else the 22 appears. One of the earliest human alphabets was the Aramaic script from around 1800 BC. From this developed the Hebrew alphabet, from which the kabbala emerged – the teaching that assigned a numerical value to every letter (1, 2, 3 to 10; 10, 20, 30 to 100; 100, 200, 300 to 400). Only 22 letters – there were no letters for vowels in Hebrew – comprise the alphabet and thus also 22 numbers. In the view of ancient peoples, this script of 'sacred symbols' contained the spirit of the whole world, which is why it is also capable of naming and encompassing everything.

Here we find a connection to the sun and its 22-year rhythms, for the sun also shines upon the whole world, and represents the whole world. Thus the number 22, the sum of the perfect numbers ten and twelve, seems to have a sun-related majesty.

23

The Ordering of Human Life

When they create a child, parents provide the chromosomes – but they do not breathe the spirit into him.

Viktor Frankl

Human male chromosomes

It happens fifty million times a second in the human body, and despite its microscopic dimension it must be among the greatest miracle that nature brings forth continually. Into the clustered chaos of the life of the nuclei of fifty million body cells, order suddenly enters. Under the microscope individual, strip-like pairs of chromosomes appear in the muddle inside the cell. As though governed by spirit hands, they migrate into the cell's central area, its equator zone, and are each held by fine threads to one cell pole. The chromosome halves release their connection to each other so that now the same genetic substance exists twice and the cell can divide into two identical cells. The threads draw the divided chromosomes to themselves, the cell ties itself off and becomes two cells in which the chromosomes loop again into a free cluster. Order is once more relinquished, the two cells start to grow and the chromosomes double for the next cell division.

'Cell division' is actually an unfortunate phrase for this process, suggesting that something already in existence is divided in the same way perhaps that a piece of bread is cut in half. It would be more correct to speak of 'cell creation' or 'cell formation'.

Every animal and plant has its typical number of chromosomes. The size of this number does not however indicate any hierarchical status in the natural kingdom. For instance, the chicken has 39 pairs of chromosomes, more than the human being (23 pairs), and the earthworm only has a few less. Nevertheless, animals and human beings seem to possess their own special number for, as soon as this alters – for example due to a complication at fertilisation – serious impairment occurs, such as in Downs Syndrome where there is a third copy of chromosome 21.

The number 23 belongs to humans but hardly any other animal (in the plant kingdom, the ash tree, for instance, also has the same number). With every cell division, and all forms of growth and reproduction, human life is microscopically ordered in line with the number 23 or, since chromosomes arise as pairs, the number 46.

What is the nature of 23 that endows it with this role in our microcosm? Number lore gives us an indication. Like five, seven or eleven, 23 is a prime number. This means that no other number is contained in it. Like these other numbers, therefore, 23 is solitary, original. If you develop a more sensitive appreciation of numbers, you start to feel that these numbers have a strongly distinctive character of their own.

Seven and 23 play a special role amongst the prime numbers, for they are followed by a larger number of divisible numbers. Thus seven is followed by eight, nine and ten – three divisible numbers – until another prime number comes with eleven. After 23, five divisible numbers first follow the prime number. Seven and 23 – and then not until 89 – are numbers that precede greater prime number gaps and can therefore be seen as being stronger in character.

Plant breeders can double the number of chromosomes in cultivated plants and thus produce bigger potatoes or apples. This is scarcely possible in animals and humans without losing fertility. Every higher organism seems to possess its distinctive number: 16 pairs in bees, 24 in apes and 47 in goldfish.

24

The All-Encompassing Number

How often have you seen these celestial bodies
shining, yet have they not found you different on
each occasion? They are always the same
however, and always have the same message:
Through our regular passage, they repeatedly say,
we determine each day and hour. Ask yourself
too, how you relate to day and hour.

Johann Wolfgang von Goethe, *Wilhelm Meister's Apprenticeship*

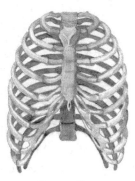

Human ribcage

We can approach the inner nature of numbers by three means. The first is through mathematics. Each number tells something about itself when we examine how it relates to other numbers. The realm of numbers reveals itself like a landscape within which every number has its place, thus showing something of its inner aspect. In addition, numbers appear in nature in distinctive, characteristic ways – in plants, animals and humans. And finally – and this is their loftiest aspect – they manifest in particular ways in culture and religion. In the case of 24, a unique number in its relationship with others, it is worth pursuing all three paths. It is the smallest number to bear within it seven other numbers as factors – the numbers 1, 2, 3, 4, 6, 8 and 12. No other number possesses such a wealth of aspects and is the vessel for so many other numbers. In this respect 24 is also related to the number seven.

As we have seen, it is the lowest number with seven factors, and it is also the biggest number where all numbers smaller than its square root are its factors. The square root of 24 is 4.899, and 1, 2, 3 and 4 are all factors.

We can grasp the special quality of 24 in counting 24 days through December to arrive at the festival of light and love. This is most pictorially expressed in the Advent calendar tradition: the last and biggest door belongs to 24. This date is, however, older than Christianity. The Christmas festival, celebrated (at midnight mass) between December 24 and 25, follows ancient sun worship which saw this day as the festival of the *Sol Invictus*, the unvanquished sun.

Just as 24 encompasses a wealth of other numbers, it also appears in us in the form of the 24 ribs of the ribcage. This encompasses our centre and contains our inner organs. However, the last two ribs are present only in vestigial form, as 'stumps', and the penultimate ones are shortened and grow together with their neighbours, as if to keep the ribcage sufficiently open.

We also see how 24 can embrace a whole in our measurement of time. Following the Babylonian tradition we divide the day into 24 hours. This was probably the quality of 24 which led the early Mosaic scholars to divide the Tenakh, the Jewish Bible, into 24 books. Starting with Genesis, the Torah is composed of five books, then come the eight books of the Prophets and finally the eleven texts extending from the Book of Joshua through the Song of Solomon to the Chronicles of Israel.

A mathematical riddle runs as follows: take the square of a prime number (bigger than three) and subtract one ($p^2 - 1$). The result is always divisible by 24. How is it possible that prime numbers always lead back to 24? Here are three examples: $5 \times 5 - 1 = 24$; $7 \times 7 - 1 = 48$ (2×24); $11 \times 11 - 1 = 120$ (5×24).

Here's the proof: $p^2 - 1$ is the same as $(p + 1) \times (p - 1)$, or in other words the product of the two neighbours of the prime number. Both must be even numbers, that is, divisible by two, and one of the two must necessarily be divisible by four since every second even number can be. The product of a number divisible by two and by four must in turn be divisible by eight. Now, one of three successive numbers one will be divisible by three. The prime number is not, so it must be either the smaller ($p - 1$) or the larger ($p + 1$). The product of the two numbers is thus not only divisible by eight but also by three – and therefore by 24.

25

In Oneself and Beyond

Let us forget there is such a thing as time,
And let us not count the days of our life!
What are centuries compared with the moment?

Friedrich Hölderlin, *Hyperion to Bellarmin*

Square numbers emanate a special power. They arise when a number is multiplied by itself, enhanced through and by itself. In arithmetical terms squaring a number is a simple process that appears in countless formulae, from Pythagoras' formula for a right-angled triangle through the law of gravitation to Einstein's famous formula of energy and mass ($E = mc^2$). But what is really happening when we multiply a number by itself? It is self-evident that one cannot multiply apples by pears. But what happens when we multiply pears by themselves? This is geometrically possible. Through multiplication with itself, a number no longer designates a series of things or steps, or in other words a linear process, for the line becomes an area. We reach a new dimension – at the expense of a loss of clarity, for an area is less tangible than a straight line. One cannot walk an area but only grasp it in thought. Something similar happens with the quality of square numbers, enhanced from three to nine, from four to 16, yet at the same time losing something of their initial, tangible clarity.

Among square numbers the squares of prime numbers play a special role, since the same thing applies to them as to the numbers they emerge from: they have no alien factors. Whereas 36, the square of six, is divisible by four or twelve and can therefore also be reached without squaring, 4, 25 and 49 relate exclusively to their base number. They are therefore 'pure' square numbers. In the usual way of recording the decimal system, 25 is distinguished from the two other square numbers 4 and 49 by retaining the figure 5, so that we can immediately recognise its 'family'. No other number is so clearly and obviously a square number as 25.

It is the smallest square number that can be shown as the sum of two other square numbers ($3^2 + 4^2 = 5^2 = 25$), a fact the ancient Egyptians were aware of and used in the construction of right angles. A rope with twelve evenly spaced knots, could be formed into a 3–4–5 triangle, thus giving an angle of exactly ninety degrees.

Twenty-five is thus the queen of square numbers and this is also probably why it has special status in the esoteric tradition, for instance in an arrangement of five times five candles at the centre of so-called magic squares.

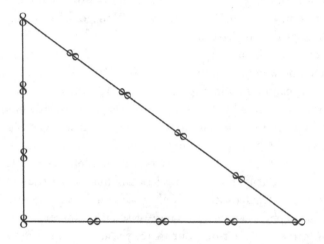

Egyptian knot triangle.

Another aspect of the power of square numbers is found, for instance, in the 25 cents as a 'quarter' (of a dollar), or the sestertius (quarter of a denarius) in the Roman Empire, or 25 years as a quarter of a century. When we arrive at 25 we are already thinking of, and prefiguring one hundred. This is why 25 years is perhaps the most significant period of time in a marriage, since this silver wedding achieves a length of time that echoes or reminds us of a century. This is not just because it is a milestone on the way to a hundred and an outstanding square number, but also because it is the sum of all the uneven, single-digit numbers: $1 + 3 + 5 + 7 + 9 = 25$.

26

The Number of Script

The word is only the body
of our inner sentience.

Philipp Otto Runge, in a letter to Pauline Bassenge, April 1803

Pierre de Fermat

The great mathematician Pierre de Fermat demonstrated that something applies to 26 which is true of no other number. Only 26 has neighbouring numbers that are a square ($25 = 5^2$) and a cube number ($27 = 3^3$). There are square numbers and cube numbers that adjoin each other such as eight and nine, with the satisfying reflection of 2^3 and 3^2. Square numbers and cube numbers can also meet in one and the same number. The smallest number this applies to is 64, which is simultaneously 4^3 and 8^2. The privilege of being both a square and a cube number belongs to all those where the number of the square is itself a cube number. After 64 (where eight is itself the cube of two) this applies to 729, which is both 9^3 and also 27^2. In the whole realm of numbers, however, there is only a single occasion when a number occupies a gap between square and cube, between, as it were, two and three dimensions. And this is 26. As in other instances, Pierre de Fermat published this fact but concealed the proof – in order to challenge other mathematicians. However the proof was so complex that his contemporaries such as John Wallis and Kenelm Digby failed to substantiate Fermat's thesis.

The 26 is thus characterised by its special position between distinctive neighbours. This allows us to find another quality of 26 that likewise relates to its location – the Latin alphabet comprising 26 letters. This most widespread of writing systems consists of 26 building blocks, to which some languages add their own accented letters.

One of the mysteries of cultural history is that 26 appears as the number of written script a millennium prior to the Latin alphabet in the four Hebrew letters YHWH, the tetragram of the Creator God and Redeemer. As in other ancient scripts, vowels were not fixed in script, due to their high status. In Judaism, following the commandment 'You shall not take the name of the Lord your

God in vain', this name YHWH cannot be uttered. Only the high priest was allowed to speak it at Yom Kippur, the highest festival. As this tradition was stopped after the destruction of Jerusalem in AD 70, the correct manner of speaking the sequence of consonants also vanished, so that it is unclear today whether we should say Yahweh or Yehovah. In accordance with ancient numerology, every letter corresponds with a numerical value. Many religious designations reveal something of their nature if one can decipher their numerical significance – which may seem surprising since the assignment occurs in a purely schematic way. In the case of YHWH, the values are ten for *yod*, twice five for *heh* and six for *wav*, giving 26. Thus 26 is the number of the inexpressible God. It is also interesting that in the biblical account, 26 generations extend from the Creation to the moment when Moses receives the script in the form of the Torah, so that something divine becomes earthly. This nature of script is expressed in the Greek word 'hieroglyph', which is a translation of the Egyptian 'sacred sign'.

27

The Number of Space

He that of greatest works is finisher
Oft does them by the weakest minister.

William Shakespeare, *All's Well that Ends Well*

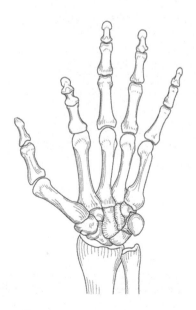

It may seem surprising, but something upon which our gaze rests most often contains the number 27. This is the human hand, whose movements are involved in so many activities, from writing or painting through to eating. Each finger has three bones, and the thumb has two, giving 14. Then there are the five metacarpal bones, of which four form the palm; and finally, oriented to the wrist, are the eight carpal bones. Thus we 'grasp' the world with 27 bones.

27 is a cube number $(3 \times 3 \times 3 = 3^3 = 27)$. But unlike 8 $(= 2^3)$ or 64 $(= 4^3)$, 27 is exclusively divisible by three, and can only be disassembled into the threefold three. So 27 is a cube number in a more absolute sense, a number of space. Here three is multiplied by itself twice.

There is an interesting anthropological consideration relating to 27 which is worth pondering. If we examine the human organism, and also the human soul, we can discover a threefold division in both physical and psychological terms. Just as the psyche can be divided into thinking, feeling and will, so this division can be discovered in the body too. The head belongs to thinking and is poor in metabolism and movement. Here consciousness dominates instead. The limbs in contrast are the site of activity and will. And between these two poles we find the trunk with its organs as the centre of feeling.

A further threefold division – and this is a fundamental trait of living things – can in turn also be found in each of these major realms. For instance, we can seek signs of will, feeling and thinking in the head. The will is located in the jaws, and this chewing apparatus is the head's limbs; feeling is in the surroundings of the nose as respiratory organ; and finally thinking is in the unmoving forehead and brain. Likewise in the trunk, we have a sphere tending towards thinking in the lungs, which are accessible to wakeful awareness, whereas the heart, as centre of feeling, is less available to consciousness; and in our digestive system, in stomach, intestine and liver, the unconscious will predominates. Finally, we find all three again in the limbs. We can, for instance, refer to heightened awareness in touching as a 'fingertip sensitivity' of things.

This little excursion into the structure of the human constitution reveals a threefold realisation of three. Life means relationship. These three times three realms are not separate from each other but are in continual interplay. In mathematical terms this gives rise to $3 \times 3 \times 3 = 3^3 = 27$ forms of interplay in both body and soul.

Like the bones in the hand, 27 is very hidden here. Where does it appear in more tangible form? In the moon, which takes 29.5 days for its changing phases. Since it is invisible for about 36 hours before and after a new moon, the most we can observe it is for 26.5 days – equating to 27 nights. This is why 27 was regarded as the number of the moon in ancient times.

The connection of 27 with spatial reality can hence be drawn: we can just about discern spatial structures on the moon, and so the moon can be considered a marker for our direct experience of physical space.

28

The Number of the Moon

Moon Night
It seemed as though the sky
Had quietly kissed the earth,
So that in blossom shimmer
She dreamt of his high worth.

Joseph von Eichendorff

28 letters of Arabic script

In ancient Greek number lore, 28 is the perfect number, for what is true of 28, few numbers can manage. The sum of its factors gives the number itself again: $1 + 2 + 4 + 7 + 14 = 28$. In mathematical terms, the factors are the content of the number, and thus 28 is identical with its content. It is therefore in complete accord with itself. Apart from the 28 this is true only of 6, 496, 8,198 then, after a large gap, 33,550,336. The Greek mathematician Euclid discovered that the formula $2^{n-1}(2^n - 1)$ allows the perfect numbers to be discovered, with $2^n - 1$ being a prime number. Thus for $n = 3$ we have: $2^{3-1}(2^3 - 1) = 2^2 \times (8 - 1) = 4 \times 7 = 28$.

To understand 28 better, it is worth studying its factors four and seven. Seven is the number of development, of time, and four designates the earth in the four directions of the compass, the four elements and the geometric picture of the square. Taken together, they signify that earth and time or development unite; or that, for example, a development leads towards the earthly realm. In addition, 28 is the sum of the first seven numbers, which underscores its strong relationship to the number seven.

The connection between earth and development points to the moon, which passes through the zodiac in almost 28 days – a period which was already regarded in ancient times as four times seven days: roughly a week from new to half moon, from half moon to full moon, and likewise from full to half moon again.

Since the earth has meanwhile moved further, it takes two extra days for the same phase of the moon to reappear. In its 28-day passage through the zodiac, the moon measures out 28 moon stations (or mansions) – special positions that were observed very carefully in many ancient cultures such as Egypt, India or China.

In Christianity too, 28 is significant. Citing an account in the Old Testament, the philosopher Albertus Magnus said that the macrocosm is composed of the mystical body of God consisting of 28 phases or types of light; and in Freiburg Cathedral there is a sequence of 28 figures from the Old and New Testament which depict the path to God.

Twenty-eight appears in human beings in two ways: pregnancy lasts an average of 267 days from fertilisation to birth. The length is usually calculated, however, from the first day of the last menstruation, giving around 280 days or forty weeks. Instead of referring to nine months, therefore, it would make more sense to count ten 28-day or four-week periods.

Twenty-eight also appears in human biography. The age of majority comes at 18 or, at the latest, 20 or 21. But the moment when the impetus of physical development ceases and one's autonomous initiative takes over as the determining factor in personal development is, experience shows, shortly before the age of thirty, and frequently around 28.

For a moon religion such as Islam, it was significant that the Arabic script has 28 letters. The famous Arab mathematician al-Biruni (973–1048) thought that this fact expresses the close connection between the cosmos and the Word of God. It is also noteworthy that the Qur'an mentions 28 prophets who preceded Mohammed.

29

The Bridge to Higher Life

Truth can stand by itself.

Thomas Jefferson

There is good reason for connecting a person's courage and honesty with the idea of 'uprightness', 'stability', 'backbone' and 'steadfastness' – phrases which all refer to the integrity of the personality. Interestingly, 29 plays a special role here. The upright stance, connected so intimately with psychological sovereignty and autonomy, is maintained anatomically by the structure of the human skeleton, along with muscles and tendons. Each leg, from the thigh bone down through shinbone and fibula to metatarsal and toes, has 29 bones which make it possible for us to stand upright. Higher up, 29 reappears in the vertebrae rising from sacrum through lumbar and thoracic vertebrae to the neck.

Whereas these vertebrae are different in number in every higher animal, all vertebrates are identical in having seven cervical vertebrae which conclude the S-shaped curve of the spine. With five conjoined vertebrae in the sacral and lumbar region (making ten), twelve thoracic vertebrae and seven cervical vertebrae, a total of 29 sculpted bone segments render the human stance possible. Most types of ape possess thirty vertebrae in addition to the tail vertebrae, but otherwise the numbers of vertebrae vary greatly in different mammals and birds. A frog has only nine, but a giant snake has over 400. The 29 human vertebrae may therefore seem random.

However, there are several mathematical phenomena which accentuate the distinctive nature of 29. It marks a threshold in the number kingdom: whereas thirty is formed of three different prime factors (two, three and five), all smaller numbers are formed only of two different numbers. For instance, twenty is composed of two and five, and 28 of two and seven. Therefore 29 is a bridge to a higher level of complexity. This higher character already applies to 29 itself: it is the first number that can be expressed as the sum of three square numbers – $2^2 + 3^2 + 4^2 = 29$.

Twenty-nine also appears in two heavenly bodies, the moon and Saturn. The moon takes 29.3 days in its passage from one full moon to the next; and Saturn takes over 29 years in its journey through the zodiac. Both planets, interestingly, mark boundaries in the planetary system. The moon delimits the sphere of earth from the further reaches of planetary space. The same is true of Saturn at a higher level, as the boundary of the realm of the classical planets observable with the naked eye. It therefore distinguishes this sphere from trans-Saturn planets that can only be seen through a telescope, and ultimately also from adjoining interstellar space. The moon not only marks a boundary to a higher planetary realm but is also capable of endowing life with shape and structure. The same is true of the 29 vertebrae, intrinsic to overcoming of a horizontal orientation natural in animals, and at the same time responsible for giving the human organism structure and stability.

30

The Great Circle

The greatest riddle of the universe is
that everything rotates.

Arthur Eddington

The number thirty appears in the thirty Goldberg variations by Johann Sebastian Bach; the initial aria presents a melody and bass which is subsequently explored in thirty different ways. The cycle, which Bach modestly called *Clavier Übung* – a piano exercise – is a pinnacle of baroque composition. The number thirty also figures in German literature in the narrative cycle *The Thirtieth Year* by Ingeborg Bachmann, and in history in the Thirty Years War from 1618 to 1648, which plunged Europe into terrible catastrophe, laid waste to whole regions but at the same time marked the birth of modern Europe. In the Middle and New Kingdom of ancient Egypt, the pharaoh endorsed his rulership after thirty years, at the so-called Sed festival.

> Thirty is an age before which certain offices cannot be held. In ancient Rome a person could only become a senator after the age of thirty, and this is still true today for holding the office of the same name in the US Congress.

Thirty corresponds to the span of each new generation – based on the idea that it takes thirty years to grow from being a child to becoming a mother or father, and that a century contains three generations.

Interestingly, the slowest of the visible planets has this rhythm: it takes Saturn nearly thirty years to complete its passage through the zodiac. Today's high life expectancy means that a human life is composed of roughly three Saturn cycles. In the esoteric tradition Saturn is seen as the planet of growth and maturity.

If we take this periodicity as a template of biography, we can discover three types of growth: up to the age of thirty the human body is developing, and this physical maturation forms the basis for psychological maturity. Up to thirty we become more intelligent whether we wish to or not! After this first third of 'soul-physical' growth there follows a period during which we can only transform into qualities of our character what we have fully grasped by our own powers. In this sense it is a period of soul-spiritual growth, for the soul only grows through knowledge and insight. The thirty years after age sixty show a kind of intensified inner growth that we might call 'spiritual-physical', for, according to many at this stage of life, physical vitality is determined primarily by spiritual flexibility and dynamism. One old person put it like this: 'If a day passes with no new ideas, things start going downhill!' White hair and a face that grows more inwardly illumined express how the spirit of the human personality can increasingly shine through the body.

After each period of thirty years, therefore, a new phase arrives. This can also be seen if we examine thirty from a mathematical perspective. Whereas all smaller numbers can be shown as the product of, at the most, two prime numbers, thirty arises from the first three prime numbers: $2 \times 3 \times 5 = 30$.

Finally, a question that can't be overlooked: why precisely thirty pieces of silver in the story of Judas? We can probably seek the reason in Jewish law of the time. If a person was killed by a domestic animal owned by someone else, the animal was stoned to death and then its owner had to pay a penalty of thirty coins. Thus the number thirty has a humiliating story in the New Testament – a plausible explanation but, as ever in religious accounts, not an exhaustive one.

A further explanation is this: that thirty itself, with the betrayal of Christ and the martyrdom that follows it, heralds the higher level of Christian existence that includes and encompasses death.

31

The Number of Mediation and Connection

Nerves are the senses' higher roots

Novalis, *Fragments*

Every year thousands of people sustain a spinal injury, often due to a fall or a traffic accident. This often leads to fracturing of one or several vertebrae. The finger-thick rope of nerves in the vertebral column is crushed or, worse, sundered – with dire consequences. Alongside the severity of the contusion, much also depends on the precise location of the lesion, since a pair of nerves exit from the spinal column between every vertebra, to innervate organs and muscles. Human beings possesses seven cervical vertebrae, twelve thoracic vertebrae, five lumbar vertebrae and five sacral vertebrae.

Before and between these 29 vertebrae, a spinal nerve detaches itself from the spinal medulla, which comprises several million nerve strands. As well as the thirty pairs of nerves that emerge from the spinal column to serve organs and muscles, there is a last spinal nerve in the rump, where a further five vertebrae are fused. There are thus 31 pairs of nerves in total which ensure a connection between organs, muscles and the brain. If the sacrum is damaged, only the last two or three spinal nerves are harmed, causing no paralysis but only impairment of digestion. Between the high lumbar vertebrae, spinal nerves migrate into legs and feet, so that an injury there can, depending on the particular vertebra damaged, impair the foot motor system, the knee or even the femoral musculature and hip movement.

Between the thoracic vertebrae lie the spinal nerves that govern trunk stability and temperature regulation, so that an injury there is far more serious and impairs independent upright sitting. Here every vertebra counts, for between the uppermost thoracic vertebrae and lower cervical vertebrae run the spinal nerves for arm, hands and fingers. The three top cervical vertebrae include the nerves responsible for head movements and breathing, so that an injury here results in grave risk to life.

Sometimes there is fortune in misfortune. A car knocking into a motorbike rider sent him flying off the city motorway in Berlin into a wood beside the road, hurling him backwards through an arm-thick tree. When a firefighter was about to lift the apparently lifeless bike rider, a paramedic called out that he must on no account move him. The paramedic had seen the broken tree and suspected a vertebral injury. In such an instance the smallest movement can twist the unprotected nerves or further crush them.

His spine was therefore first set firmly in place, and then a blanket was slipped underneath his body in order to lift him without jolting. The subsequent diagnosis showed that this caution saved him from spending the rest of his life in a wheelchair.

The brain gains its connection to the body through the 31 pairs of spinal nerves. These nerves are the 31-membered spiritual connection between the brain as centre and the organs and muscles as periphery.

Are there other suggestions that 31 is the number of connection? It is the eleventh prime number and thus stands at the threshold of the full circle of the first twelve prime numbers. It is also the sum of the first five doubled numbers, $1 + 2 + 4 + 8 + 16 = 31$; and the sum of the first three powers of five: $5^0 + 5^1 + 5^2 = 1 + 5 + 25 = 31$. Thus 31 bears various numerical processes within it. Perhaps it is this quality that renders 31 capable of forming a totality of interconnections so as to mediate between organ and brain processes.

**Floris
Books**

For news on all our **latest books,**
and to receive **exclusive discounts,**
join our mailing list at:

florisbooks.co.uk

Plus subscribers get a FREE book
with every online order!

We will never pass your details to anyone else.

CPSIA information can be obtained
at www.ICGtesting.com
Printed in the USA
JSHW020715201120
9678JS00014B/76

9 781782 506645